THE PRISONER

Robert Muchamore

Hodder
Children's
Books

THE PRISONER – KEY LOCATIONS

GREAT BRITAIN

LONDON ★

RAF Bexhill ●

ENGLISH CHANNEL

Dieppe ◉

Beauvais ●

PARIS ★

- - - BORDER

RIVER

SEA / LAKE

◉ KEY LOCATIONS

GERMAN OCCUPIED

FRANCE

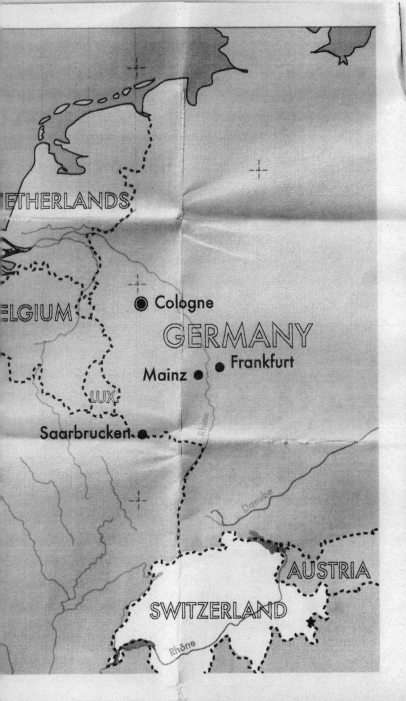

Part One

May–June 1942

> 'This war will be over before America is
> ready to begin fighting.'
> Adolf Hitler, 1942

In early 1941, Britain stood alone against a Nazi empire that controlled most of Europe. This changed on 22 June when Hitler betrayed his ally Stalin and began a massive invasion of Russia. Five months later, Japan launched a surprise attack on the United States at Pearl Harbour and America became the last great power to enter the war.

As 1942 began, the Second World War had become a global conflict, with the Axis powers led by Germany, Italy and Japan lined up against the Allies: led by Britain, Russia and the United States.

On paper, the Allies were stronger. They could muster more men and produce enough weapons to crush the Axis. But the USA was ill-prepared for war, while Germany was fully militarised and controlled huge sections of Russian territory.

To win against the Allies, Hitler had to beat Russia before America reached full fighting strength. As he threw all his military resources into the war against Russia, back in Germany over ten million prisoners toiled, producing the food, fuel and weapons needed for the largest land battles the world has ever seen.

This army of workers comprised captured soldiers, criminals, communists, Jews and other groups persecuted by the Nazis, plus forced labourers drawn from occupied countries such as Poland and France. From pensioners to teenagers, these slaves lived on meagre rations, with poor sanitation and limited safety equipment, while under constant threat of punishment by brutal guards.

CHAPTER ONE

Frankfurt, Germany, May 1942

The sky was the colour of slate as Marc Kilgour crossed a damp gangplank on to the *Oper*. The old steamer had spent three decades taking passengers along the River Main before fire crippled her. After years sulking at dockside, layered with rust and soot, war had brought her second life as a prison hulk.

Oper was bedded in a remote wharf east of Frankfurt's centre and only floated off her muddy berth on the highest tides. All windows above deck had been boarded and the passenger seating ripped out and replaced with stacks of narrow bunks.

Marc had lived aboard for eight months; enough time that the fourteen-year-old barely noticed the stench of bodies and cigarettes, as he walked down a gangway between bunks that was barely wider than his shoulders. Almost all the other men were out at work, leaving behind sweat-soaked straw mattresses and graffiti etched into pine bed slats.

A man groaned for attention as Marc passed. To get off work you had to be seriously ill and while Marc didn't know him, he'd heard how the big Pole had crushed his hand while coupling freight wagons, then picked up a nasty infection that was working up his arm.

The words came in a half-delirious strain of Polish. The man wanted water, or maybe a cigarette, but he was crazed with pain and Marc upped his pace, wary of getting involved.

The timber stairs that led below *Oper*'s main deck still bore the scars of fire. Charcoal black rungs creaked underfoot as Marc's hands slid down a shrivelled stair rail. The stench below deck was denser because the air got less chance to move.

All three light bulbs in the passageway had burned out. Marc felt his way, counting eight steps, passing a foul-smelling toilet, then stepping through a narrow door. A mouse scuttled as he entered the wedge-shaped room. Mice were no bother, but the rabbit-sized water rats Marc occasionally encountered freaked him out.

Marc had no watch, but guessed he had an hour before his five roommates returned from twelve-hour shifts in the dockyard. He groped in the dark, finding the Y-shaped twig they used to prop open their oblong porthole.

Fresh air was a privilege – not many cabins below deck had them. The light revealed two racks of three bunks against opposing walls, with a metre of floor space between them. Upturned crates made chairs and a wooden tea chest served as a table.

One of Marc's predecessors had fixed up a shelf, but everyone kept their mess tins and any other possessions tucked under straw mattresses: theft was rampant and it was riskier feeling around a bunk than stealing from an open shelf.

Marc dug into his trouser pockets, pulled out two small, rough-skinned apples and let them rest on the table. He'd swiped them from the Reich Labour Administration (RLA) office earlier. He was easily hungry enough to eat them, but the six cabin mates always shared food.

They were a decent bunch who looked out for each other. Sometimes Marc would score fruit, bread, or even cake left over after a meeting in the admin offices. His cabin mates who worked in the dockyard or train depot occasionally got their mitts into cargos of food.

The mouse resurfaced, scuttling along a bed frame and out the door as Marc climbed on to his bunk. It was on the third tier of four. With half a metre to the next bunk, it was impossible to sit up.

After sweeping some dead bugs off his blanket, Marc unlaced his wrecked boots. His feet had grown and his only pair of socks was stained dark red where his heels and toes rubbed raw. But the itching under Marc's shirt bothered him more than his bloody feet.

The straw mattress rustled as he unbuttoned his shirt. Marc was naturally stocky, but prisoner rations had been poor – particularly during the cold months between December and February – and he'd lost all the fat over his rib cage. He scratched at a couple of new flea bites as he aligned his hairy armpit with the light coming through the porthole.

Marc combed his fingertips through the mass of sweaty hairs. Sometimes you had to hunt the louse making you itch, but today a whole family had hatched in one go. He squinted as he picked half-a-dozen sesame-seed-sized body lice out of his armpit, squishing each one against the wall for a satisfying crunchy sound.

The next phase of battle was a hunt for nits – trying to pick out sticky eggs before they hatched. With so many bodies packed on the boat, with most prisoners only having one set of clothes and no proper washing facilities, body lice, fleas and bed bugs were inescapable.

Picking out bugs always depressed Marc. It was hard being far from everyone he knew, being hungry and being forced to work, but the bugs and filth were worst because they meant he didn't even control the most intimate parts of his own body.

When Marc had done his best with the lice, he turned on to his back and stared at the mildewing wooden slats of the top bunk, less than an elbow's length from the tip of his nose. He was fiercely hungry and his mind drifted, but his hand slipped under his straw mattress and he smiled warily as he felt a piece of green card in his grubby hand.

Just touching it scared him. He'd been trying to escape since arriving in Frankfurt ten months earlier, and removing it from the administration office was a risk. If everything worked out, a card like this would be his ticket out of Germany. But if he got caught, it could just as easily become his death warrant.

*

Marc was no ordinary prisoner. Reich Labour Administration records said he was Marc Hortefeux, a fifteen-year-old French citizen from Lorient, sentenced for smuggling black-market food, who'd volunteered for agricultural labour in Germany.

In reality he was Marc Kilgour, a fourteen-year-old from Beauvais near Paris. Orphaned shortly after birth, Marc had escaped to Britain after the German invasion of France two years earlier. He'd then been among the first batch of young agents trained to work undercover for an espionage group known as CHERUB.

Imprisoned by the Gestapo during a sabotage mission, Marc had been forced to kill a fellow inmate who'd bullied him. He'd faced a death sentence for murder, but a French prison commandant took pity and agreed to commute Marc's sentence, provided he volunteered for five years' labour service in Germany.

To qualify for this programme Marc needed to be fifteen years old. The commandant ensured that Marc's prison records were lost, and a replacement set drawn up with a false age and giving him a longer sentence. The next afternoon Marc had boarded a train to Frankfurt and he'd been here ever since.

*

'Sleep, eh?' sixteen-year-old Laurent shouted, as he slapped Marc gently across the chest. 'Lazy bastard.'

Marc's eyes opened as he shot up, almost thumping his head on the bunk above. More than two hundred inmates had finished a shift, and as well as the sound and smell of his roommates, the Oper's cabins and passageways had come alive with shouts and clomping boots.

'Just resting my eyes,' Marc said, as his mouth stretched into a yawn. 'Reading documents is a strain.'

Laurent shook his head wryly as he unbuttoned a shirt coated in grey dust. 'Poor little eyeballs,' he laughed. 'All we had to do today was haul bags of cement.'

Laurent had been on German rations long enough to get skinny, but he still had the solid jaw and vast fists of someone you wouldn't pick a fight with.

'Pen-pusher,' Marcel added, as he squatted on to the bunk below Marc's, peeling back his shirt to inspect skin scoured by the heavy sacks.

Marcel's words were harsh, but the tone was warm. Marc's cabin mates were envious that his ability to speak German had earned him an admin job, but none of them seemed to resent his good fortune.

Marc rolled on to his side, trying not to inhale a grey haze as four sweating lads stripped off clothes thick with cement dust.

'There's a couple of apples on the table,' Marc said.

'We'll get fat on them tiny buggers,' Marcel replied.

Marcel was a joker. Only fourteen, his crime was to lead cheers in a Rouen cinema when a newsreel showed the aftermath of a British air-raid in Cologne. The Gestapo officer two rows back didn't see the funny side and Marcel found himself riding to Frankfurt, minus two front teeth.

'Grub's up,' Richard – the last of Marc's cabin mates to arrive – shouted, as he stepped in holding a battered roasting tin. It held two loaves of black bread[1] and a tall metal jug, with steam rising off a thin, orangish soup.

Richard was a Belgian, fifteen but with tiny, sad

---

[1] Black bread – a coarse, near-black loaf, traditionally eaten by European peasants who couldn't afford refined white flour.

eyes and a genteel shuffle that made him seem old. As he placed the roasting tin on the table, his roommates dived under their mattresses to grab spoons and mess tins.

'If I divide this, you'd better not moan,' Richard said.

'I'll divide if you don't want to,' Marcel said eagerly, lunging towards the loaves.

The food on the tray was dinner and breakfast for six hungry teenagers, and the lads would fight over every crumb. Marc was lucky to have roommates who'd played fair, even during the harshest winter rationing. There were plenty of other cabins where bullies ripped off weaker inmates' food.

'Marcel, you mess around with that bread and I'll slam your head in the porthole,' Laurent said, firmly. 'Richard's always fair, leave it to him.'

'Yeah, Marcel,' a lad called Vincent added. 'Especially seeing as you've spent half the day picking bugs out of your crotch.'

There was some laughter, but it was also an uncomfortable reminder of the squalor they all lived in.

Prisoners weren't allowed knives, so Richard broke the bread into six fairly even clumps with his filthy hands, then began ladling the soup into six differently shaped mess tins. Hungry eyes tracked every move of the ladle.

'Give me more!' Vincent said. 'Marc's is way deeper.'

'His tin's round, yours is square,' Richard said. 'You both got four spoonfuls.'

Vincent folded his arms and pouted. 'I always get screwed.'

Laurent took a mildly intimidating step towards Vincent. 'He's spooning it all out the same.'

Out of his cabin mates, Vincent was the only one Marc didn't care for. He wasn't a bad guy, but he was always having digs about stuff, and that grinds you down when there's six of you living on top of each other.

'Have my tin if you think there's more,' Marc said irritably.

Before Vincent could answer, the tension was broken by a body thumping down hard on the main deck above their heads.

'Fight,' Marcel said, staring up at the ceiling as cheers and shouts echoed down from the cramped main deck.

As the ruckus continued, the six lads grabbed their mess tins, and settled on the wooden crates, or propped awkwardly on the edge of the lowest bunks. Marc eyed his soup and poked his spoon in, spotting a few identifiable chunks of vegetable and strings of horse meat, amidst thin gruel made from swede and potato.

The hungry boys dispensed with their soup in under a minute, then licked out the tins. Their black bread was days old and made slower eating. Marc stuffed a crusty end into his cheek and began softening it with his back teeth as he lay on his bunk.

'I'll cut each apple into six pieces,' Richard said, as he pulled his identity disc over his head.

The metal ovals were stamped with each prisoner's number and worn around the neck. Rubbing your disc against stone gave a sharp edge, which was no substitute for a proper knife but better than nothing at all.

'Five pieces,' Marc said. 'I had mine earlier.'

'And the rest,' Vincent sneered. 'I bet you scoff all kinds of shit over in that office.'

'It's easier cutting something into six,' Richard complained.

'Next time I'll eat it myself and save the moaning,' Marc said, as Richard cut into the first apple.

The apples were bitter and the lads screwed up their faces, but nobody complained because they appreciated the risk Marc had taken, smuggling food out of the administration office when he could have scoffed the lot.

As the six boys settled on their bunks, mouths stuffed with bread, a shout came down the passageway from the top of the stairs.

'Raus!'

It was the first word of German every prisoner learned. It meant *out*, but the guards used it as a kind of rebuke: *get out of bed, move out, get ready*. If the guards were in a mood and you stood close enough, *raus* would be accompanied by a flying boot or ball of spit.

The six lads in Marc's cabin all launched curses as a pair of German guards clomped down the stairs. Prisoners got called off the boat for all kinds of reasons: roll calls, searches for contraband, delousing.

'Boots on,' a guard shouted in bad French as he leaned through the doorway. It was Sivertsen, a squat, fair-haired Dane, who'd volunteered for the German army. The Russian shrapnel lodged in Sivertsen's back left him with a severe tremor in his arm. The inmates regarded him as a bit of a joke.

'What's going on?' Laurent asked, as he swung his legs out over the edge of the top bunk.

'Obey,' Sivertsen shouted. 'No questions.'

Laurent got his answer from another guard, who spoke better French and was explaining to the lads in the next cabin that a train had derailed and needed to be unloaded before it could be safely lifted back on the tracks.

As Sivertsen turned to leave, he noticed that Marc hadn't shifted from his bunk.

'Did you not hear?' Sivertsen roared, as he shoved past Richard and closed on Marc's bunk.

Marc spoke in German. 'I work for the administration office, not in the goods yard.'

Prisoners pulled all kinds of tricks to get out of work. Sivertsen didn't believe Marc and placed a hand on the baton hanging from his belt. 'This *is* an emergency. If I tell you to get up and work, you get up and work.'

'You need to speak to Commandant Vogel if you want to use me,' Marc said, then with a cheeky smile: 'But he'll have gone home for the day.'

At the same moment, Marcel crept up behind Sivertsen and made a loud quacking sound in his ear.

The Dane pirouetted with his stick. The blow glanced off Marcel's elbow as he dived on to his bunk and pulled up his mattress as a shield. Marc and the other lads started laughing.

'What's this?' a senior guard named Fischer roared through the doorway. 'Why is this taking so long?'

Sivertsen was a joke, but Fischer scared everyone. He was a Great War veteran on the wrong side of sixty, but a lifetime hauling cargo in the docks had kept him tough and he had a reputation for stomping inmates who talked back.

'We're all getting ready, sir,' Laurent said.

Fischer gave fellow guard Sivertsen a contemptuous look. 'Are you in control here, officer?'

'Yes, sir,' Sivertsen said, anxious not to look weak in front of his boss. 'Lad here says he works for the commandant.'

Marc was about to explain, but Fischer yanked him

towards the edge of his bunk and clamped a hand around his throat.

'If your feet aren't in your boots in three seconds, I'll have you shitting blood for a month. All clear, inmate?'

'Yes, boss,' Marc croaked.

CHAPTER TWO

Marc was among sixty workers, crunching through the gravel between railway tracks, driven to move fast by four guards at each end of the column.

'Made you work for once, eh?' Alain said, as he gave Marc a jab between the shoulder blades.

While Marc's cabin mates didn't hold his admin job against him, the same couldn't be said for some of the other lads. Alain was a thug in his late teens. He bunked with thirteen others in a big cabin next to Marc's.

They were a rough crew who Marc tried to avoid. If he couldn't avoid them, he'd stick close to Laurent, who'd stand up for him. But Marc's undersized boots were killing him. He'd fallen back in the darkness and lost sight of his friend.

'Searching for lover boy?' Alain asked, as he gave Marc another dig.

'Get off me,' Marc spat.

The next jab was harder and made Marc stumble. He turned sharply and hissed.

'What you gonna do?' Alain teased. 'Take a swing, see what you get.'

Marc reckoned he could take Alain, even though he was older. But Alain's goons would wade in if Marc got the upper hand. Even if he beat that lot, the price of victory would be a kicking from the guards.

The prisoners bunched and stopped as a man on horseback trotted on to the tracks up ahead and started addressing the guards.

'No more dawdling, you lazy pigs,' the rider shouted pompously, as the column moved off at a brisk jog. 'Double time!'

Jogging pace shouldn't have bothered young men, but physical work and hunger wears you down. The workers had heavy legs and it was easy to stumble on a railway sleeper.

As the prisoners crunched past, the man on horseback took swipes with his riding crop, knocking one lad into another and catching a straggler nastily across the face.

'Idle French scum,' he roared, as he trotted along the tracks at the rear of the column, ready to swipe anyone who fell back. 'The main line out of Central Station is backed up. You'll work hard and fast or I'll have the lot of you flogged!'

Marc almost tumbled as he straddled a lad who'd fallen heavily, but he was less tired than the lads with manual jobs. He steadily advanced towards the front of the column, losing tormentor Alain as they jogged three kilometres along darkened tracks.

The scene of the derailment emerged as they came around a lazy bend in the track and reached multiple tracks heading out of Frankfurt Central Station. Searchlights from an anti-aircraft battery had been swung around, illuminating a dozen sets of parallel rails with a low viaduct crossing them.

A goods train passing over this viaduct had skipped the track. Several wagons had jack-knifed, shedding wagonloads of lumber and bails of flax on to the main line below.

As Marc paused to catch breath, he noted French markings on the engine pulling the derailed cargo train. He'd never heard of a derailed train during his first months in Frankfurt, but there now seemed to be at least one a week and he wondered if they were being sabotaged.

The idea that the resistance was still active cheered Marc slightly, but the prisoners standing around him were more practically minded: to them bails and logs spilled over train tracks represented nothing but hours of gruelling labour.

'No sleep tonight,' Laurent said, shaking his head.

Marc was relieved to have Laurent back in his sights. As the two lads exchanged smiles, a trio of Germans on horseback gathered a few metres ahead, discussing plans with breathless foot guards who'd had to run alongside their prisoners.

In the background, a scrawny prisoner was being dragged out of the darkness by Fischer. It was the boy Marc had almost tripped over, and as well the gash in his forehead he had a bloody lip where Fischer had punched him in the mouth. The injustice of it reminded Marc of how powerless prisoners were and made him boil with anger.

'Teams of ten,' Fischer roared, as he shoved the injured kid away with a kick up the backside. 'Take the logs away from the main line and up the embankment. The faster you work, the sooner you sleep.'

There were already a couple of passenger trains blocked in behind the logs, including a Berlin express with irritated passengers staring out the windows.

The logs were an assortment of rough shapes heading for a sawmill. With no ropes or chains, the gangs of ten struggled to get any kind of grip on the logs, and even when they doubled to teams of twenty it was slow and brutal, taking the huge logs across the railway tracks, then up a steep muddy embankment.

A crowd of railway officials, police and soldiers gathered, but none of them helped the prisoners. There were more than fifty logs to move and by the time Marc was on his third trip his arms and shoulders were badly strained. He had grazes up his arms and splinters in his hands.

Once they'd cleared logs lying on their own, the teams had to deal with logs that had settled in dangerously unstable piles. Nobody was surprised when a scream went up and a Dutchman's ribs got crushed under two tonnes of wood. The prisoners got him out quickly, but he was barely breathing.

'Do that again, it was funny!' a drunken soldier shouted from above.

Marc was exhausted and used the rescue of the Dutchman to squat on a rail, studying his splinters and clutching aching sides. Laurent was alongside him.

'Your hands are soft,' Laurent explained, as he showed Marc dark calloused palms.

'German bastards,' Marc said angrily, as a couple of men carried the semi-conscious Dutchman up the embankment. 'Why can't they use horses to move the logs?'

Laurent laughed. 'Why damage a nice horse when you've got us?'

'I'm not laughing,' Marc said. 'What if it's you or

me trapped under those logs next time?'

'I wonder if this is all there's ever gonna be,' Laurent said. 'If Germany wins, Hitler's not just gonna say *bye-bye, all you lovely prisoners can go home*, is he?'

'They won't win,' Marc said determinedly, as he thought of something Commander Henderson had once told him. 'America has over half of the world's industry. They're on our side now and they can produce more planes and weapons than the rest of the world put together.'

Laurent stared at the empty black sky. 'That may be so, but I've seen none of 'em in this neighbourhood.'

Marc looked around to make sure nobody was in earshot. 'We could escape,' he said quietly.

'Dream on,' Laurent said. 'The soldiers up on the viaduct would see and shoot us both in the back.'

Marc felt for the piece of green card in his pocket. He wanted to show Laurent, but it was too risky to pull it out here, and he didn't want to spoil it with his bloody hand.

'Not right now,' Marc said softly. 'I've spent a lot of time in the RLA office. Prisoners get transferred all the time. I've worked out the system. I think . . .'

Before Marc could finish, Laurent tugged him to his feet. 'Look sharp, someone's coming.'

The man riding towards them wore the grand uniform of the German transport police. As the hoofs of his speckled grey horse crunched in the gravel, the boys braced themselves for a riding crop across the face for sitting down on the job, but the German's voice was surprisingly warm.

'You two look tired,' he said, speaking the stilted French that rich German boys learn at school. 'Do you want to help move some of the flax bails?'

'Yes, boss,' Laurent said grudgingly.

'Thank you, sir,' Marc added.

Laurent scowled at Marc as they hurried across the tracks. 'Never *thank* them, you crawler.'

'I'm not crawling,' Marc said. 'But I'm more likely to survive 'til sunrise carrying bails than hauling massive logs.'

'He's feeling guilty 'cos one of us got half killed,' Laurent said. 'Give him an hour and he'll be back to lashing out.'

'What the hell is flax anyway?' Marc asked, as they started walking.

Laurent shrugged. 'They use it to make cloth, I think.'

While the wood was spilled over the central tracks, the flax bails had tumbled from a pair of flat-bed carriages at the rear of the train. They were the first on the scene, and Marc immediately saw plusses and minuses.

The big plus was that the bails of flax strands were light compared to giant logs and weren't going to kill you. On the downside, the bails stank of mould and jumped with fleas; while getting them off the rails involved carrying them through overgrown nettles alongside the tracks and then up a muddy, steeply sloped embankment.

With no close supervision, Marc and Laurent cleared a path by stamping down nettles, before carrying the first bails up the embankment. They wanted the job to last so that they didn't get sent back to deal with logs, but they couldn't work too slow in case it set off a guard.

By their fifth run, Marc and Laurent were flea-bitten and mud-caked. Three more lads, including Alain, were freed up to carry bails when horses and chains finally arrived to take on the dangerous task of dragging logs out of unstable stacks.

'What a surprise, Marc gets the cushy job again,' Alain said, when he recognised his nemesis halfway up the embankment with a huge bail balanced across his back. 'How come so many Germans love you? Are you a snitch, a queer, or both?'

Marc turned slightly, which wasn't easy on a muddy slope with a weight on your back. 'Funnily enough, Laurent and I were just talking about you,' Marc said. 'We were wondering if your mother still made good money whoring herself out to the Gestapo?'

The other two new arrivals half laughed and half gasped as Alain's eyes bulged.

'You don't speak about my mother,' Alain shouted

as he started a muddy charge towards Marc. 'You're dead, Hortefeux.'

Laurent shouted from the top of the embankment, 'You touch my friend, you touch me.'

Marc felt confident with Laurent on his side. When Alain got close, he let go of the bail, which was heavy enough to sweep Alain's feet away. Alain crashed in the trackside gravel at the base of the embankment, then sat up to be met by Marc's muddy boot connecting with the bridge of his nose.

'Like that?' Marc roared, as he pushed a mound of flax out of the way.

He landed knees-first on Alain's chest, and his fist gave Alain a bloody mouth to match his bloody nose.

From the top of the embankment Laurent could see Fischer and another guard running towards them.

'Marc, back off,' Laurent shouted.

But Fischer had his rifle aimed at Marc's chest before he could make three paces.

'Freeze,' Fischer shouted. 'Face me.'

Marc turned with his hands in a surrender position.

'On your knees,' Fischer shouted. 'Hands on head.'

As Marc's knees squelched into the mud, Fischer lunged with his rifle butt. It was the kind of blow that could fracture a skull, but luckily the big German misjudged and the swing only glanced Marc's brow.

'You dare give me trouble twice in one night?' Fischer

shouted, as Marc splashed into the mud with blood gushing from a five-centimetre cut above his eye. 'You're in my black book now, Hortefeux, and that's *not* a good place to be.'

CHAPTER THREE

It was gone 2 a.m. when Marc got back to his bunk. He was woken again at six, by guards booting doors and shouting. Marc could feel the swollen lump and sticky blood over his eye, but had no mirror to see it properly.

'How's it look?' he asked.

Laurent peered down from the bunk above, with what remained of last night's bread ration softening inside his right cheek. 'You need a stitch, but you ain't gonna get one. At least you can keep it clean in your job.'

Marc had a banging headache. His muscles ached from lifting logs and he needed to pee, but he avoided the scrum in the filthy washroom and fought through bodies and cigarette smoke to be one of the first workers off the boat.

The sun was barely up and it was drizzling again. Not

that Marc minded: The moisture was refreshing after a stifling night aboard the *Oper*.

'Good morning, Herr Osterhagen,' Marc said in his politest German, when he reached the guard hut at the bottom of the gangplank.

Osterhagen was decent. The young guard rarely searched anyone and only acted tough when he had to put on a show for Fischer. He had no obvious ailments, which was curious because healthy German men fought on the front lines, while prisoners were guarded by the old and unfit.

Marc's theory was that Osterhagen had friends in high places.

'Looks like you suffered last night,' Osterhagen said, as he reached into the guard hut and grabbed a filthy jacket with KG^2 painted on the back in large red letters.

'Courtesy of Herr Fischer,' Marc replied.

'He's a nasty one,' Osterhagen said, opening a wire gate and letting Marc out as he pulled the jacket up his arms. 'Keep out of his path.'

When he'd first arrived in Frankfurt, Marc was surprised that all but the highest-security prisoners were allowed to walk to work with nothing more than a set of initials on their back. Occasionally someone

---

[2] KG – an abbreviation of Kriegsgefangener, which means 'war prisoner'.

took advantage, but Marc had never heard of anyone getting away.

Frankfurt was deep inside Germany, it was impossible to travel without documentation, few prisoners spoke German, you needed ration stamps for food and there were security checks everywhere. As a further deterrent, recaptured inmates could expect a beating, followed by reassignment to a coal mine in Silesia.

Once the gate was out of sight, Marc dived into an alleyway and took a long piss against a wall. After that he set off on a six-minute walk along the riverbank towards the Reich Labour Administration offices in the Großmarkthalle.

Although it was early, there were plenty of people about, including the unpleasant but familiar sight of young toughs in Hitler Youth uniform. This security detail was all fifteen or sixteen years old and looked ludicrous in white knee socks, short shorts and camel coloured shirts with swastika armbands.

'Where are you going?' the tallest asked, as they surrounded Marc. One standing behind him drew a long wooden baton and tapped it menacingly in his palm.

'Großmarkthalle,' Marc said, before tutting. 'Just like the last five times you've stopped me.'

'This little Frenchie has a *bad* attitude,' the leader said, and the others all laughed. 'Show me your prisoner disc.'

Marc undid a button on his shirt and pulled out the

disc. One of the Hitler Youths took out a pencil and wrote his number in a notebook.

'You really stink,' one lad said. 'Don't they teach French boys to wash?'

'You can pass,' the leader said as he stepped out of Marc's path. 'But you make our dockside look a mess, we'll punish you one of these days.'

This was no idle threat. Hitler Youths dressed like boy scouts, but they were all indoctrinated with racist ideas and the authorities turned a blind eye when they chose to demonstrate their Nazi spirit by intimidating and beating up foreigners.

As Marc stepped away, the quartet broke into caustic laughter.

'He was trembling,' one said triumphantly. 'Skinny French weed.'

Marc would have loved to have a go back, but that was a losing ticket, so he chewed his lip and contented himself with the thought that they'd all get sent to the Eastern Front as soon as they turned seventeen and hopefully end up on the wrong end of a Russian tank.

Großmarkthalle was a vast market hall built along the riverbank. The length of three football pitches, it had been built as a wholesale vegetable market, but the building along with surrounding railway sidings and docks had been turned into a transport hub for the military. Großmarkthalle moved troops and prisoners, along with the chemicals and manufactured goods produced in Frankfurt factories.

A familiar guard let Marc through a gate and he passed into an echoing hall, vaulted with huge concrete ribs thirty metres above the floor. Reich Labour Administration was based in a six-storey office building at the opposite end and Marc walked the length of the hall, dodging tired-looking prisoners rolling tyres and barrels towards a waiting goods train.

Close to the offices, over a hundred prisoners squatted in a wooden pen, guarded by three Gestapo men with rifles and Alsatian dogs. Six-pointed stars sewn on smart clothes marked the prisoners out as Jews. They were neatly shaved and well fed, because unlike foreign prisoners the Jews were locals who received extra food from non-Jewish friends and female relatives living in the city.

The Reich Labour Administration offices were on the fifth and sixth floor and the lift was only for Germans. Marc had to clank his way up ten flights of metal stairs. The office was mainly staffed by typists and file clerks, who worked in a brightly-lit pool area, with windows overlooking the River Main to one side and the interior of the market hall on the other.

Before eight the only people around were usually Marc, a German caretaker and two Ukrainian cleaning girls. So Marc was surprised to hear Commandant

Vogel already in his office, shouting furiously into his telephone.

Marc didn't want to get on the wrong end of the Commandant in a temper, so he grabbed prisoner record cards from a tray marked *To be Filed* and hurried off. His aim was to quickly wash his face and bloodied hands in the bathroom and then hide out in the archive room upstairs filing the cards.

But Marc only got a couple of steps before Vogel charged to his doorway, with a telephone receiver at his cheek and the cord stretched across the room behind him.

'Here,' Vogel shouted, as he waved his arm. 'Put those down, get over here.'

The prettier of the two young cleaning girls gave Marc a smile and a *good morning* nod as he waited in the office doorway, listening to Commandant Vogel deal with an angry factory boss.

'Bloody Gestapo!' Vogel shouted, after slamming the receiver down. 'Have you seen those Jews downstairs?'

'Yes, sir,' Marc said.

'Gestapo brought them in for immediate deportation to Poland. I had enough of a headache when they took all the Jews doing excavation work. But this batch are skilled men: chemists, doctors, engineers. The factory bosses are up in arms. They're irreplaceable!'

Vogel thumped his desk before raising his voice to a

new level of fury. 'I'm supposed to be in charge of the RLA for the whole of this area, but the Gestapo chops my legs off. How can we win the war when skilled workers get sent to do farm work in Poland?'

Marc thought it was best not to mention that fact that *he* didn't want Germany to win the war and mumbled, 'I don't know, sir.'

'I can't get a phone call through to our headquarters in Berlin. I need you to run to the main post office and send this telegram.'

Marc took the handwritten note from his boss. 'I'll be there as soon as they open the doors, sir.'

'What?' Vogel spluttered, as he glanced at his watch then backed up to his chair. 'Apologies, I keep forgetting that it's early. I've been kept up half the night by factory bosses begging to get their Jews back. It feels like lunch, but it's not even breakfast.'

'I can go down and fetch you something to eat from the canteen, sir.'

Vogel smiled. 'Yes, I'll give you my military ration card.'

As Vogel reached into his desk drawer, his swivel chair bashed his desk, knocking a mound of papers to the floor. Marc bent down to pick them up, but a painful spasm erupted in his back.

Vogel was startled. 'Are you sick?'

'Just stiff,' Marc explained, as he restacked the papers

and placed them back on Vogel's desk. 'They had us out half the night moving logs off the main line.'

'You work for *me*,' Vogel said, as his posture straightened with irritation. 'I need you fit, not stiff and covered in blood. Why didn't you tell them that you worked for me?'

'People try to get out of extra work,' Marc explained. 'I did say, but the guard didn't believe me.'

'Then you should have asked to speak to the area supervisor.'

'Fischer *is* the area supervisor, sir,' Marc said.

'That thug,' Vogel said contemptuously, as he half stood up and studied the shape of the wound above Marc's eye. 'That was his rifle butt, I'll bet.'

Marc knew there might be consequences to snitching on a hard case like Fischer, but if he lied Vogel would only ask for another name. Marc's mistake had been letting Fischer's name slip in the first place.

'It was, sir.'

'Right,' Vogel said furiously as he picked up his phone and raised it halfway to his mouth. 'You clean up that face. I want you at the doors of the post office when it opens. Get that telegram off to Berlin the second they open, then take my ration card. I just want coffee. But you use it to get yourself something decent to eat.'

When you're hungry food is the only word you hear. Marc beamed at the thought of German military rations:

processed cheese, bread and maybe even a hard-boiled egg if he was lucky.

As Marc stepped towards the door, Vogel aimed his hand towards the private bathroom at the rear of his office.

'There's soap and hot water,' Vogel said. 'But *don't* touch my towel. I can live without body lice or scabies.'

Laurent's comment about never grovelling to Germans flashed through Marc's head, but soap and hot water was almost as big a treat as the offer of decent food.

'Thank you so much, sir,' Marc said, as Vogel brought the phone the rest of the way up to his mouth.

'Osterhagen?' Vogel told the handset authoritatively. 'Is Herr Fischer still on duty . . . ? Excellent. Tell him he's required in my office when his shift finishes.'

CHAPTER FOUR

Marc felt better with the blood scrubbed off his face and military rations in his belly, though feeling good was still some way off. As a minimum it would have required better fitting boots and a good delousing to stop his constant itching.

After sending Commandant Vogel's telegram, Marc translated in a short meeting with a French volunteer worker, who wanted to return home to look after his three children following the death of his wife.

When that was done, Marc stepped into a secretarial pool that was now filled with the chatter of typewriters. He went around all the secretaries asking if they had any prisoner records to file and took the resulting mound of cards up to the sixth floor.

The layout was identical to the offices below, but rows

of file cabinets were lined up in front of the windows, blocking most of the light. The room was stuffy, so Marc propped a fire door open with a waste paper basket.

The door led out on to a balcony overlooking the River Main. It was a bright day, and from up here the world looked serene: barges steaming along the river, tall buildings in the city centre with their tops engulfed in the smog rising off dozens of chimneys in the industrial belt beyond.

Marc didn't linger outdoors, because while most of the younger secretaries pitied him, the office was run by two elderly sisters who treated the grubby French teenager in their office with the same regard that they might show a diseased rat.

Marc's stack of cards was twenty centimetres high. Every foreign worker and prisoner in Germany had a Reich Labour Administration record card. The rows of file cabinets surrounding Marc contained cards for more than three-hundred-thousand men and women controlled by the RLA's Frankfurt district.

There were green cards for French prisoners, pink for Poles, blue for Jews, yellow for Russians and Ukrainians and so on. Cards had fingerprints and photos, names, prisoner/worker numbers and details of past and current work assignments.

Cards were filed by prisoner number. Marc's job was to replace cards that had been taken out by secretaries, mostly to write on details of new work assignments, hospitalisations, or disciplinary measures.

He began with sixty Russian prisoners, who'd all been quarantined following an outbreak of typhus in their barracks. The next batch were all deaths. If a German or West European prisoner died, their card was marked with a red X and sent to Berlin, where relatives would be sent notification.

If they were East European or Russian, Marc logged the dead prisoner's name and number in a register and the cards were put in a stack for incineration.

One of the many personal items that prisoners didn't receive was toilet paper, and Marc often tore these surplus cards into quarters and slipped them into a pile behind one of the cabinets. When he had a good stack, he'd smuggle them back to the *Oper*.

Marc would give the papers to his cabin mates – Laurent always excused himself to the toilet with the phrase, *off to wipe my arse on a dead Russian* – or traded them with dock labourers on the upper deck, who'd swap a three-centimetre stack of cardboard squares for carrots or a couple of medium-sized potatoes.

While Marc filed, a young secretary named Ursula came in and started pulling cards for the Jews waiting downstairs in the market hall. She wasn't exactly beautiful, but glancing at her when she bent over to put something in a bottom drawer took Marc's mind off the tedium of filing and the bugs partying under his grubby shirt and trousers.

Commandant Vogel caught Marc staring when he stepped in, but instead of rebuking him Vogel took his own longing gaze up Ursula's skirt and gave Marc a nod as if to say, *not bad*.

'I've just had the pleasure of Herr Fischer's company,' Vogel announced. 'You'll not be sent out to do manual labour again and Fischer will let the other guards know the same thing. He's also got orders to find you some better boots and clothes. I know it's not your fault, but some of the girls downstairs find your odour offensive.'

'I try, sir,' Marc said. 'But we never get any soap. I've only got one set of clothes and there's nowhere to wash them.'

Marc appreciated the fact that Vogel looked out for him, but it was a double-edged sword: most prisoners were older and bigger than Marc, there was already a lot of resentment about his cushy job and getting better boots and clothes could easily set off a thug like Alain.

Vogel failed to notice Marc's wariness because he'd spotted what Ursula was doing and threw a fit.

'Who gave you permission to pull those records?'

'The Standartenfuhrer called from the Gestapo offices downstairs,' Ursula explained. 'He wants all their cards transferred to his office. There are already Jewish women at the perimeter making a fuss and asking to see their men. They're adding cattle wagons to the next train east so that they're out of our hair before it turns into a scene.'

'Damned Gestapo,' Vogel snapped.

'Shall I stop, sir?' Ursula asked.

'I suppose not,' Vogel said bitterly, as he ground a palm against his forehead with frustration.

Ursula stood rigid, cleared her throat and spoke nervously. 'Herr Commandant, I don't wish to speak out of turn. I know you're determined to keep up factory production, but you've voiced your opinions about the Gestapo and the Standartenfuhrer rather loudly. He's a powerful man. I'm not the only one who worries that being so vocal might not be good for you.'

Vogel broke into a big smile. 'It's sweet that you care, Ursula.'

'You're a considerate boss,' Ursula said, slightly embarrassed. 'All us girls enjoy working with you.'

This little conversation intrigued Marc: the more he dealt with the Nazi state, the more he saw how their racist policies and tangled bureaucracy worked against them. They were desperate for electrical engineers and vehicle mechanics, but Marc had seen men of both professions shipped off to mining districts because they were Russian. Jewish doctors got stripped of their medical licences and were sent to Poland, while German troops died on the front lines because the army lacked good medics. It was all quite mad.

'When you've drawn the files, give them to Marc,' Vogel told Ursula. 'He can deliver them to the Standartenfuhrer with my compliments.'

*

'Good evening, Your Majesty,' Herr Fischer purred, doffing his cap sarcastically as he let Marc through the dockside gate. 'There's a present on your bunk. I hope it's to sir's liking.'

'Right,' Marc said awkwardly.

'One other thing,' Fischer added, when Marc was halfway up the gangplank. 'Should you ever fall out with your chums in high places, I *won't* forget that you gave the Commandant my name when he asked for it.'

Marc shuddered, though luckily it was dark enough that Fischer didn't get the satisfaction of seeing it.

A late meeting between Vogel and the French foreman at a new engineering plant had kept Marc out past nine and the other prisoners were all back from work before him. To get to his bunk below deck, Marc had to interrupt several card games being played in the narrow gangway and got sworn at for his trouble.

Marc jolted when he saw Alain and his boys at the bottom of the stairs playing dice. Alain didn't speak, but even in the gloom below deck his busted nose and swollen lips weren't pretty. Marc had wanted to escape since the day he'd reached Frankfurt, but making two powerful enemies in one day paid off any lingering doubts.

'Evening, boys,' Marc said, as he stepped into his cabin.

Most of his cabin mates were half asleep, though Richard and Louis – a ginger fifteen-year-old serving twenty years for distributing communist literature – were squinting over a two-week-old French newspaper that had come their way via a convoy of French coal barges.

Marc's share of the evening rations was congealing on the upturned tea chest, but the package on his bed was more interesting. Wrapped in brown paper and tied with string, its presence was novel enough to attract everyone's eyes as soon as Marc touched it.

'What you got, golden boy?' Laurent asked.

Laurent was Marc's closest friend, but his sarcastic tone was another sign of how preferential treatment caused bitterness.

Undoing a shoelace knot and peeling back paper revealed a small, dark-brown man's suit and a pair of recently-resoled leather shoes. They looked too big, but that would be less painful than Marc's present boots which were two sizes too small.

'The commandant must have really put the frighteners on old Fischer,' Marcel laughed. 'I'd have loved to see his face.'

As Marc unravelled the jacket, he was surprised to see

a Star of David patch sewn on the breast pocket. His brain instantly connected dots between the brown suit and a small Jewish man he'd seen squatting on his luggage in the market hall that morning.

Marcel laughed. 'I'd unpick that star before the Hitler Youth stop you on the street.'

Two thoughts sprang into Marc's head. First, was a mental image of his suit's former owner, riding a train to Poland in his underwear. Secondly, for Fischer to get one of the Jews' suits, he must have visited the Gestapo office downstairs after being told off, and Marc doubted that Fischer would have had good things to say about Commandant Vogel.

'You'll look like a little Jew businessman in that,' Marcel said, laughing. 'You just need the big hooter and thick glasses.'

'Smart threads will earn you an arse kicking too,' Vincent, the eternal optimist, added.

'I'll only wear it for work,' Marc said. 'Maybe even leave it at the office so it doesn't pick up too many bugs.'

Marc grabbed his mess tin and started to eat, acting hungrier than he really was. Your stomach shrinks when you're always hungry and he didn't want his friends knowing that he'd scored bread, synthetic jam, cheese and scrambled egg at the Großmarkthalle earlier on.

'There's something I have to talk about,' Marc said, as

his spoon scraped out his tin. 'Come up this end, I don't want anyone in the passageway hearing.'

Marcel made everything a joke and found Marc's grave tone amusing. 'You didn't finally get your tiny dick inside a German secretary, did you?'

As his five cabin mates shifted towards him, Marc pulled the green card from his trouser pocket.

'This is a Reich Labour Administration record card,' he began. 'Every foreigner working in Germany has one. Ours are all stored in the massive filing system above the office where I work.'

'So what?' Vincent asked.

'I've been studying the system,' Marc explained. 'Which documents you need for which process: transfers, hospitalisation, repatriation. Which signatures you need, which stamps go in which boxes. I think I can get paperwork that will get us all back to France.'

'Escape?' Richard asked, recoiling at the thought.

Marc shook his head. '*Escape* is when the guards miss you at the next roll call and the Gestapo hunt you down. This is different. French prisoners and volunteer labourers sometimes get sent home on compassionate grounds. They get issued with release papers and travel warrants for their home town in France.'

'I see,' Laurent gasped, as he jumped down off his bunk to examine the green card. 'But how do you get all this paperwork?'

'I can do a signature that looks enough like Commandant Vogel's to fool the girls in the office. The blank forms are lying around all over, and I can use the rubber stamps and do my own sneaky bits of typing, because I usually arrive at the office before anyone except the cleaners.'

'You say you *think* you've worked it out,' Vincent noted. 'What if you're wrong?'

'I've done most of the work already,' Marc explained. 'I've got two sets of paperwork sorted for all of us. It's prepared and hidden between two filing cabinets back at RLA headquarters.'

'Two sets?' Laurent asked.

Marc explained. 'Repatriations are rare. If we all try getting out of here with paperwork sending us to France, the guards will smell a rat. So we'll leave this barracks with orders transferring us to a new work detail in Cologne. But when our train gets to Bonn, we dump that set of paperwork and produce release documents and travel warrants enabling us to board a train to France.'

'This is my third barracks in Germany,' Louis said. 'You *can't* just hop off the train. Prisoners always travel long distance with armed escorts, or locked in cattle trucks.'

'I've got that covered,' Marc said. 'We're prisoners when we leave here. When we get to the Frankfurt Central, the paperwork will say we're volunteer workers, with travel permits for regular trains.'

'And if someone at the station sees the switch, or recognises us?' Laurent asked.

'No plan is perfect,' Marc said, sounding frustrated. 'It's a risk to go, but so's staying here. In this filth we could easily pick up TB or typhus. They can transfer us to a mine or some factory where you choke from all the chemicals. Horses get more respect than the Germans show us.'

'That Dutchman who went under the logs died,' Marcel said.

'That could have been any one of us,' Marc said.

'There's no need to get tetchy, Marc,' Laurent said. 'We respect your plan, but we've got a right to ask questions before putting our heads in the noose.'

'Fine,' Marc said, understanding Laurent's point but still irritated by how negative everyone sounded. 'Ask away.'

'What happens when we don't arrive in Cologne?' Marcel asked.

'Normally the alarm would go up,' Marc explained. 'Headquarters posts cards when workers transfer to another district. Cards arrive at the local RLA Headquarters a day or two after the prisoners. But in our case, our cards will simply vanish into thin air. Or more accurately, I'll bring them back here and destroy them before we leave.'

'So you're saying we get back to France, free men?' Laurent asked. 'All our paperwork is in order and nobody will even be looking for us?'

Marc nodded. 'Take your release documents to the local identification office and you'll be issued with an up-to-date French identity card and ration book under whatever name you choose to give me for your paperwork.'

'Sounds too good to be true,' Vincent said. 'If your system is *so* perfect, how come nobody else has figured this out?'

Marc laughed. 'Maybe someone has. How would you know unless they were dumb enough to brag about it when they got back to France?'

'Speaking of France,' Richard said, 'In case you haven't noticed, I'm not French.'

'I looked into that,' Marc replied. 'Problem is, prisoner returns to Belgium and Vichy France[3] run on a different system. Release requests get assessed in Brussels or Vichy. So it's France or nothing, I'm afraid.'

'Count me out,' Vincent said, shaking his head knowingly. 'Schemes like this always go wrong.'

Marc shrugged. 'I'm not forcing anyone, but I have to know who's in before I leave in the morning.'

---

[3] Vichy France – a southern, mainly rural, area of France that was not occupied by German troops. Although it was officially an independent nation, Vichy France was run by the pro-Nazi puppet government.

'I've got no family or anything in France,' Richard said. 'It's not *so* bad here on the docks. If we're caught, we'll end up down some Polish coal mine.'

Marc looked anxiously towards Laurent, keen for someone to say yes.

'In our situation you've got to balance risks,' Laurent said. 'It sounds like Marc has thought things through and I'd rather chance this than waste away here.'

'Agreed,' Marcel added. 'I'm in.'

Louis nodded thoughtfully. 'Last January I got so hungry I thought we'd all die. I'd rather throw the dice now than risk another winter.'

'So that's four of us,' Marc said, not displeased with that result. Vincent was gutless and he'd never expected him to say yes, though Richard's decision was disappointing.

'So when does it all happen?' Marcel asked. 'I've gotta say goodbye to all my lady friends.'

Laurent burst out laughing. 'Only lady friends you've got will come with you inside your diseased brain.'

Espionage training had taught Marc that the longer a plan gestates, the more chance there is of someone blabbing.

'I need to type the dates and add this week's RLA authorisation stamp on all the paperwork,' Marc said. 'I'll get that done tomorrow, and put in a transport request for an escort to the station. If that works, an RLA truck will arrive at the gate the day after tomorrow and take us to the station for the seven-eighteen train to Cologne.'

'Paris by the weekend,' Laurent said cheerfully. 'Hell yeah!'

CHAPTER FIVE

Although Marc had been planning the escape for two months, it only felt real now that he'd told his cabin mates about it.

He dated and stamped all the cards and travel warrants before any of the secretaries arrived. Then he checked and rechecked: *Had he used the right stamps on the travel warrants? Were all his German spellings correct? Had he properly submitted the request form to the transit office on the third floor?*

It was a quiet day at the office. Commandant Vogel received a reply to the telegram he'd sent to Berlin and seemed unusually subdued. Marc was usually happy with a slow pace, but today it gave him too much time to think of what might go wrong.

Marc felt a twinge of fondness when Vogel dismissed

him for the day, just after six o'clock.

'I like the new suit,' Vogel said. 'See you bright and early tomorrow.'

'Have a good night, sir,' Marc said, knowing that he'd never see him again if all went well.

After putting on his prisoner jacket, Marc snuck up to the filing area on the sixth floor. There were a couple of secretaries doing their end-of-day filing, but he had no bother slipping his hand between two cabinets and retrieving a big envelope stuffed with the papers he needed for the escape.

There was always a chance of being searched on the gate when you arrived back at the *Oper*. Marc hoped to see Osterhagen sitting in the guard hut, but it was the Dane, Sivertsen.

The always-keen Sivertsen would search you if you arrived alone, so Marc lurked in a doorway for twenty minutes until two trucks arrived, carrying mud-encrusted prisoners who'd spent the day digging foundations on a construction project outside of the city.

Marc merged into the crowd at the gate, successfully shielding himself behind a big man, while Sivertsen patted down a moustachioed Belgian with his shaky hand.

Marc's head was full of things that could go wrong between now and morning: an unscheduled prisoner count, delousing, random search, one of the other lads coming back from work with an injury.

Marc braved the evil-smelling toilet, which was actually just sloping boards mounted over a shit-spattered hole cut into the *Oper*'s hull. Two vast rats glared as he ran water into his mess tin.

Back in his cabin, Marc lay on his bunk and began destroying four green prisoner cards, tearing them into centimetre squares, dropping them in the tin and squeezing them until they'd turned to mushy green pulp.

The evening was routine, except there was more to talk about than usual. Laurent said Richard could have his mattress after he'd gone. It was the only one filled with cotton rather than straw. They took turns guessing their chances of getting away: Louis guessed sixty-five per cent, Laurent eighty, Marcel was the most pessimistic with fifty-fifty. Marc refused to put a figure on it.

The optimism level improved when Marc gave his three fellow escapers their documents. Richard couldn't fail to be impressed, seeing it all typed in German on official stationery, with proper stamps and Marc's nifty forgeries of Commandant Vogel's signature.

'Kind of wish I'd said yes now,' Richard said ruefully, as he inspected Laurent's Bonn-to-Paris travel warrant. 'We've been a good crew.'

Marc got his few possessions together and tied them in a bundle using a piece of rag, and the string from the package Fischer had brought down the day before. He hugged the small bundle to his chest when everyone settled in for the night, but he was way too tense for sleep.

Marc was paranoid that he'd put a tick where a cross should be, or missed a security question, or forgotten some document entirely. He'd grown fond of his bunk mates and wanted to help them escape, but their lives depended on him getting every detail right, and part of Marc wished that he'd just used the scheme to vanish on his own.

*

Osterhagen came below just after 5 a.m., waking half the boat by shouting out four prisoner numbers, followed by, 'Transfer orders. Feet! On your feet!'

Prisoners were never notified of transfers, in case it encouraged them to hide, escape, or switch identities, so Marc acted mystified when Osterhagen stepped into the room.

'What do you mean, transfer?' Laurent asked.

'Is it local?' Marc asked.

'Why us four?' Laurent added.

'You think they tell me anything?' Osterhagen scoffed. 'Get moving.'

Marc pushed his feet into his new shoes, threw his cloth bundle over his back and quickly hugged Vincent and Richard, before leading the quartet outside.

It was sunrise as they walked down *Oper*'s gangplank under orange-tinted smog. Their ride to the station was by open-backed horse-drawn cart, staffed by an armed Reich Labour Administration transport officer, with a woman at the reins.

Before they passed through the gate, Osterhagen had a procedure to follow, checking that the prisoner identity discs matched the information requested on the transfer documents and crossing their names off the *Oper*'s register so that the boys wouldn't be missed at the next head count.

It had been dark below deck and Osterhagen's face twisted awkwardly when he recognised Marc.

'Fischer got bollocked over you,' Osterhagen told him. 'Says nothing happens to you without him or Vogel saying so.'

Marc and Laurent exchanged a wary glance, as the transport officer stepped up to the gateway.

'Do you know where they're going?' Osterhagen asked.

'Orders to take them to Central Station came through from the office late last night,' the transport officer explained. 'Quite weird actually: transfer requests *always* come through a few days ahead of time.'

Marc gulped. A minor mistake before he was even out of the gate. Was it the only one?

'Sorry to hold you up, but I've got to call my boss,' Osterhagen told the transport official, as he pointed to Marc. 'This kid works for Commandant Vogel. We've got special orders not to let him out for manual work. It might apply to transfers too.'

Marc's heart bounded as Osterhagen flipped open a metal box mounted by the gate and pulled out a telephone handset. With his fate dangling, Marc gleaned all he could from Osterhagen's end of the conversation: Fischer was off duty and whoever Osterhagen was talking to seemed to be pushing the decision back on to the cautious young guard.

'Can't stand about here,' the transport officer told Osterhagen, as he tapped the face of his pocket watch. 'I've got to collect transfer prisoners from two other barracks, then take new arrivals out to Florstadt.'

Osterhagen shuffled his feet and wrung his hands, clearly unused to making decisions.

'Take the other three,' he said finally. 'I might get bollocked if a transfer is delayed, but that's nothing to what Fischer might do to me if I let the boy leave when I shouldn't have.'

Marc felt sick as he said a quick goodbye, before Laurent, Marcel and Louis walked through the gate and climbed up on the cart. They understood the escape plan and he'd given them their own documents in case they got split up on a crowded train, but seeing his three friends' shocked faces as the cart rode off without him was crushing.

'So what now?' Marc asked, locking his fingers together to stop his hands trembling.

'They're going to try and find Fischer,' Osterhagen said. 'Failing that, they'll wait until the commandant comes on duty.'

As Marc peered back towards the *Oper*, the lump in his throat made it hard to breathe. He thought about overpowering Osterhagen and making a run for the station. He might manage, but even if he made it on to the train the alarm would be raised before he got to Bonn and he'd blow the other three's covers in the process.

'Have a sit inside the hut,' Osterhagen suggested. 'There's a flask of coffee if you're thirsty. The phone could ring any minute. *Someone* must know about this transfer.'

CHAPTER SIX

Predictably, Vogel quashed the transfer order and Marc walked to the RLA office in time to start his regular shift. Prisoners got picked for random transfers all the time, so there was no investigation and Vogel joked that he'd saved his young messenger and translator from having to do real work in Cologne.

'When your card is returned, I'll make sure it's marked, *No transfers*,' Vogel said, when he called Marc in to collect a batch of telegrams.

'Good idea, sir,' Marc said, faking enthusiasm.

Marc briefly considered what would happen when the card that he'd pulped and dropped out of the *Oper*'s porthole didn't show up. But cards were lost or misfiled all the time and they just wrote out new ones.

He got more worried when Vogel stood up from his

desk and circled behind, like an interrogator.

'Blond hair, blue eyes, strong body, upright posture,' Vogel said mysteriously, as his hand rested on Marc's shoulder. 'You have what my boss, Reichsfuhrer Himmler[4], would call *desirable racial characteristics*. There's even a program where boys like you get Germanised.'

Marc looked baffled, which made Vogel laugh.

'Germanised: it's all part of our racial policies. Jews get booted off to some mud patch in Poland. Foreign boys and girls who fit the image of tough, blond-haired Germans get rebadged and adopted by German families, or sent for training in Hitler Youth Camps.'

Marc thought the idea that kids could become German based solely on the way they looked was barking, but was too diplomatic to say so.

'I can look into it if you'd like me to,' Vogel said. 'No bugs and I'm sure the food's better.'

As one door closes, another opens, Marc thought. But an obvious downside was that healthy German boys not much older than himself got handed uniforms, guns and one-way tickets to the Russian Front.

'Well?' Vogel asked. 'It's not like you to stand there gawping.'

---

[4] Heinrich Himmler – German Interior Minister, head of the SS, Gestapo, Reich Labour Administration and many other departments. Himmler was regarded as the second most powerful Nazi after Hitler.

'I'm . . .' Marc said, before tailing off. 'It's just the last thing I expected you to say, sir. I thought you were keen to keep me here, working for you.'

'I'm sure I can find another messenger boy,' Vogel said, as he handed Marc the telegrams. 'Promise me you'll have a serious think about it and let me know within the next couple of days.'

'Yes,' Marc said. 'I will . . . Of course I will.'

Marc's head was all over the place as he put on his prisoner coat and jogged across town. The Germanisation idea was weird, but better food and a lice-free existence had undeniable appeal, even if he couldn't exactly see himself in dinky little Hitler Youth shorts and a swastika armband.

The post office queue was huge and Marc watched the clock as the line crept forward. All being well, Laurent, Marcel and Louis would be at Bonn Station by now, with half an hour until their connecting train to Paris.

Either that or they were in a cell, looking at a whole heap of trouble that would lead back to him.

*

Marc had to work late with one of the secretaries, translating German instructions to a new piece of machinery for French workers. His appearance in his cabin on the *Oper* just after eight gave Richard and Vincent a jolt.

'That explains why I got served three portions in the food line,' Richard said.

'There's bread left, but we ate your share of the soup,' Vincent added. 'How far'd you get? Are the others coming back?'

'I got to the end of the gangplank,' Marc said peevishly, as he peered into the empty soup bowl. 'I've not heard anything about the other three, but if it's gone to plan, they'll be on home turf by now.'

Marc pushed a ball of bread into his mouth, but as he stepped up to his bunk he noticed his mattress had been switched for a ripped one with a strong whiff of urine.

'Who took it?' Marc gasped. 'Why'd you let them in here?'

'Alain heard you guys got transferred,' Richard said. 'Some of his boys came in.'

'Can't fight thirteen of 'em,' Vincent said. 'And you'd better watch your back now Laurent's not covering it.'

'Tell me something I don't know already,' Marc said, then flew back in shock, making a gagging noise.

As Marc straightened the mattress, he'd revealed a hole. Inside hundreds of maggots wriggled within the rotting corpse of a mouse.

'Christ,' Marc said, as he threw the straw mattress on to the floor. 'I'd rather sleep on bare slats.'

'Alain's gonna talk to Fischer,' Vincent said. 'Their cabin's crowded, some of his boys want to move in here.'

'Us six made a good set-up,' Richard said forlornly. 'Laurent kept us safe. Now . . .'

Vincent gave Marc an accusing look and finished Richard's thought. 'Now it's all screwed, thanks to *you*.'

Marc felt guilty, but also pissed off because real mates don't kick you when you're down. The swastika and shorts looked more attractive by the minute.

As Marc tried to work out whether it was better to sleep on bare boards, or if he had to find some way to clean the worst filth and stink off his new mattress, two boys from Alain's cabin swaggered in with their stuff.

'Fischer gave us the all clear,' the bigger of the two announced as he threw his mattress up on to Laurent's old bunk. 'Hope that's OK. Tough if it's not, 'cos it's happening.'

The other lad got all excited when he saw Marc. 'Thought you were transferred?'

Before Marc could answer, the bigger lad shoved Marc back against the porthole and shouted. 'Alain, get in here!'

Marc caught his captor with a palm to the face and broke free, but ran straight into Alain as he ducked through the doorway.

'Hello, old pal!' Alain said cheerfully, as he grabbed Marc by his collar and bashed his head against the side of a bunk. 'I thought I'd missed you.'

Marc broke Alain's grip and landed a good punch in

his kidney, but the space between bunks was barely a metre wide. With three older lads boxing Marc in, he was quickly overwhelmed with the one lad twisting his arm up behind his back and Alain doubling him up with a brutal kick in the stomach.

'Remember kicking me in the face?' Alain shouted, as he kneed Marc again.

Marc groaned as Alain grabbed his legs, leaving him suspended agonisingly by his twisted arm. His head got thrown against both sides of the passageway as more of Alain's crew poured out of their cabin.

'Mess that little punk up!' someone shouted eagerly.

Marc ended up thrown against the wall under the stairs, with the taste of blood in his mouth and boots flying in from all sides.

'You'll kill him,' Richard shouted, earning himself a slap and a firm warning to butt out.

Marc lay in a shaking ball when the kicking stopped, but it wasn't over. Alain knelt across Marc's chest with a home-made knuckleduster wrapped around his fist. Marc felt sure he'd die as a powerful blow smashed into his mouth.

Savage cheers and whoops came from the lads around Marc as Alain pulled back for another shot.

'Finish the little queer off,' someone shouted, as Marc gagged on his own blood.

Alain struck Marc over the right eye, exactly where

Fischer's rifle had opened him up two nights earlier. Blood spurted as Marc's head snapped to the right. He tried yelling, but the pain froze his whole body.

Then everything went black.

CHAPTER SEVEN

Time passed in a blur as Marc drifted in and out of consciousness

He felt Osterhagen's fingers down his throat as he lay flat out on the wharf, choking on blood. Then he was on a prisoner ward, with mattresses on the floor and an insect tickling his wrist. Patients coughed and spewed in buckets.

Marc ached all over and passed out from the pain if he tried to move. His ribs were strapped. He could only see out of one eye and when he touched the blind side he felt a tennis ball where his eyelid should have been.

One time – it seemed like morning – a nurse told Marc that Commandant Vogel had been to visit while he was unconscious. Then Marc came round in paradise. A clean, light ward, and real beds with German children in them. A nurse spoon-fed him stewed apple, but the lump over his eye was even bigger.

Marc was only vaguely aware of an argument, a sense that he'd been sent somewhere he shouldn't. All his dreams seemed to be nightmares of burning, drowning, being bombed or being kicked.

The next time he regained consciousness he'd gone from paradise to a straw mat in a filthy ward, full of cigarette smoke and men speaking Russian. A nurse swore when he wet the mattress and she left him laying in warm piss that gradually soaked through his hospital gown up to the back of his neck.

Stoypa, a Russian lad, started helping him sit up to use a bed pan. They spoke different languages, and Marc thought Stoypa was a nurse until he sat by himself and saw his helper squatting on the corner of the next mattress, hacking up snot.

Stoypa disappeared before Marc had a chance to thank him. He began hobbling around, and got his sense of time back: knowing when it was day or night, or when the food was due. Hungry men heal slowly, so the Germans gave hospitalised prisoners better rations than they got in the camps. After another week, Marc could breathe OK and most bruises had almost gone.

Marc had been deloused and all his body hair shaved to get rid of the bugs while he was unconscious, but instead of being itch free he'd picked up a rash from lying in his own urine. But the real problem remained the stubborn swelling covering his right eye. It distorted his face so that one side of his mouth twisted into a crooked sneer.

Each morning a nurse stabbed the swelling with a needle and drew out pus. But that only relieved the pressure under badly stretched skin. Finally, there was a showdown between a Polish doctor, who said the swelling above Marc's eye needed time to heal, and a German hospital administrator with a strict quota for getting her patients back to work.

The German won the argument and Marc was sent for surgery. He'd imagined pain relief, bright lights and clean walls, but anaesthetic barely existed for non-Germans. Marc got told to sit on a chair in a gloomy basement room and tilt his head back.

With leather straps around hands and ankles, and two bear-like Russian prisoners holding him down, Marc watched a scalpel and screamed as the blade cut deep and twisted inside the swelling. Marc thought he'd pass out from the pain, but briefly got vision from both eyes.

It was a relief knowing that he wasn't blind beneath the swelling, but soon both eyes were filled with blood as the surgeon cut away strips of skin, then used a sterilised tea-spoon to scoop out rancid-smelling pus and clotted blood.

When the straps were off, Marc sobbed with pain as the Russians lifted him into the world's oldest wheelchair and took him back to his floor mat. The Polish doctor pushed a nurse aside and dressed the wound himself.

'It's not always a bad thing to open a wound up,' the Pole explained, in immaculate French. 'It can speed the healing process. But in a less than pristine environment, there's great risk of secondary infection. So keep dry until a good scab forms, and however much it itches, *don't* poke about under the dressing with your grubby fingers.'

*

Three days later, Marc was sent packing with a jar of antiseptic ointment and a roll of used but allegedly sterile bandage. He spent five hours in a hallway waiting for a transport official. The wound over his eye was still sore, but he had good vision in both eyes and his brain was back in order.

From a date on a newspaper in the waiting area, Marc worked out that it was thirty-five days since Alain had punched him out. He was scared of going back to the *Oper*, so it was a relief when the driver said they were taking him to Großmarkthalle first.

Marc hadn't forgotten the whole Germanisation thing, and although he was disappointed that Commandant Vogel hadn't been back for a second visit, he hoped the offer remained open, and that Vogel could protect him from Alain and his gang.

It might seem insane for a prisoner to spend a month in hospital only to be sent back where he'd almost been killed in the first place, but Marc knew he was nothing more than a card in a file cabinet to the Reich Labour Administration. He'd seen enough prisoner death reports to know that stuff like that happened all the time.

It was a warm day and Marc almost felt euphoric on the back of the cart, breathing outdoor air, seeing women in food queues and kids with black soles dangling fishing lines off the side of a bridge.

Marc realised that a month in bed had weakened his legs as he climbed the ten flights up to the RLA's fifth-floor office. The transport official left him in the doorway, and the first person he saw was Ursula, the secretary.

'You look different with your head shaved,' she said, giving Marc a wary smile before tugging him into an alcove hidden by a wooden coat stand.

'This cut's all the rage at the hospital,' Marc said.

But Ursula was anxious. 'The new commandant went round asking questions about you,' she began.

Marc felt like he'd been punched. '*New* commandant?'

'It's been Commandant Eiffel for two weeks now. You didn't know?'

'How would I know?' Marc asked. 'Although I did wonder, because I thought Vogel might visit me again.'

'Herr Vogel put his neck on the line to get you a bed in the children's hospital,' Ursula said.

Marc's eyebrows shot up. 'Was that why he got the boot?'

Ursula shook her head. 'He'd made enemies with the Gestapo. Complained to Berlin about moving out those skilled Jews. They've transferred him to take charge of a labour district in Poland. A promotion, allegedly, but he's Frankfurt born and bred, so he wasn't keen to leave.'

'Hortefeux,' a woman barked. It was one of the two battleaxe sisters who ran the office. An imperious stab of her pointing finger sent Ursula meekly back to her typewriter.

'I hope it works out,' Ursula whispered.

Marc waited for ages outside the commandant's office. He'd developed a theory: the longer someone makes you wait, the more of a dick they are.

Vogel's name on the door had been scratched off the frosted glass, but Eiffel's had yet to replace it. Marc was surprised by the slim female figure moving about inside.

'Hortefeux,' she said, when she eventually asked Marc in.

Eiffel was stern, catlike and chain-smoked through a long ivory cigarette holder. She wore sombre civilian clothing, but with a swastika armband. A framed photo of Hitler had replaced the Mayor of Frankfurt on the wall behind the desk.

'Sit,' she said, before switching to decent French. 'My predecessor spoke highly of you. Commandant Vogel even left me a letter, recommending that I keep you on.'

Marc nodded eagerly. 'I'm sure I'd be useful, madam.'

'Was Herr Vogel's trust in you well placed?' Eiffel asked, raising one eyebrow as she flicked her cigarette end, clumsily missing an ashtray.

Marc tried to hide his discomfort. 'What do you mean?'

'All departments send prisoner transfer logs to Reich Labour Headquarters in Berlin,' Eiffel explained. 'Every movement is logged and our system is robust. It's rare that the numbers don't tally.

'But shortly after I began here, I received a letter asking me to investigate an anomaly with three prisoners. They were transferred to Cologne, but never arrived.'

Marc gulped, Eiffel smiled. She liked making him squirm.

'I couldn't get any information on the prisoners, because their records had vanished from the card index – to which you had frequent and easy access. However, a transport official remembered a slightly unusual transfer request, and Osterhagen recalled that *you* tried to leave with three prisoners bound for Cologne. Apparently you were stopped at the gate, but your three friends vanished into thin air.'

Marc thought about lying, but Eiffel had clearly investigated thoroughly. He was worried not just for himself, but because he could potentially be tortured into revealing the new identities and locations of his three friends.

'I miss my family,' Marc said meekly, hoping Eiffel would take pity. 'I just wanted to go home.'

Eiffel shrugged disinterestedly. 'I'm sure the Gestapo Security Office would be intrigued by all these details. Fortunately for you, I have no desire for a large-scale Gestapo investigation of my department. I've reconciled the prisoner numbers with Berlin. The matter is closed and you'll be assigned to a new job where you'll have no access to prisoner records.'

Marc didn't want to seem cocky, but couldn't completely disguise a relieved smile. 'Thank you, madam. Has my new work assignment been selected?'

'A senior guard mentioned that he has communication difficulties and could use someone like you who speaks French and German.'

Marc's heart plunged. 'Which guard?' he asked, though he was sure he knew already.

'Herr Fischer will be here to deal with you shortly,' Eiffel said.

CHAPTER EIGHT

Marc got a strong-armed shove, making him stumble back towards a chair in Fischer's office.

'Sit,' Fischer barked.

Calling it an office was a stretch: the wooden hut had a hammock strung across, an old door laid on trestles for a desk, empty cans and beer bottles stacked to the ceiling and a smell like wet dog.

As the thuggish former dock labourer reached up to screw an electric cooking ring into the light socket, Marc noticed a photo of his new tormentor in his prime: bare chested, muscular tattooed arms leaning cockily on the ropes of a wrestling ring.

'Ma said I was sick in the head 'cos Old Fischer used to mess with cats,' Fischer said, as he ran water into a saucepan, using a standpipe poking through the hut's wooden floor. 'I'd throw knives at 'em. Or grabbed the little bastards and slit their guts open.'

As Fischer chuckled to himself, Marc didn't know where to look or what to say. He felt uncomfortable, not just because he'd been released into the custody of a nutter, but because the hospital had badly shrunk his brown suit when they'd boil-washed it to kill off all the bugs.

'You French kept Old Fischer prisoner for two years in the last war. You think it's bad here? You should have seen how you treated us.'

Fischer worked around a tin of tomatoes with a can opener and began tipping them down his throat.

'Vogel sent Alain off to punishment camp,' Fischer said, as he held the can out towards Marc. 'Tom toms?'

No prisoner ever turned down food, but as Marc reached for the can, Fischer snatched it, then gobbed a big mouthful of chewed-up tomato into Marc's face.

'I hold my grudges,' Fischer said, grunting with laughter. 'Alain may be gone, but there's still plenty of his mates on the *Oper*. I'll put you in with 'em if you muck me about. Got it?'

'Yes, sir,' Marc said, as he wiped tomato juice and spit on to his sleeve.

'Old Fischer's in charge of the *Oper*, the main prison barrack behind Großmarkthalle and three other prison boats. Now wherever you go, there's always prisoners with a racket. Prisoner knows how to get extra food. Prisoner with gold hidden in his mattress. Your job is to find 'em and come tell me all about it.'

Marc looked stunned.

'I thought you liked flapping your trap, snitching to your boss?' Fischer teased. 'You snitched me to Vogel well enough, didn't you? Earned me and the other guards a right bollocking.'

'Prisoners who snitch get their throats cut,' Marc said weakly.

'Best be careful then,' Fischer said, with a laugh. 'But keep information coming my way, 'cos if you're no use to Old Fischer . . .'

The guard finished his sentence by swiping his finger across Marc's throat. As Marc sat there trying to think up a plan, Fischer opened another tin of tomatoes. This time he let Marc dip his fingers in and take a couple.

Marc scoffed the bitter tomatoes so fast that juice ran over his wrist into the cuff of his shirt. Then Fischer yanked his arm and sadistically bent back his fingers.

'You'll do what *I* say, when *I* say it.'

As Marc writhed off his chair and hit the floor, a small glass jar rolled out from the inside pocket of his suit. Fischer snapped it up and roared.

'Yoghurt!' he shouted. 'Haven't seen that since before the war. Did you steal it from the hospital?'

Marc flinched, expecting a boot in the gut, but Fischer

unscrewed the cap, dipped in two fingers and sucked the creamy liquid off his fingertips.

The reaction was explosive as the foul-tasting substance burned Fischer's throat and tongue.

'Christ,' Fischer roared, banging his fist on the desktop, then swirling juice from the tomato tin around his mouth to clear the taste. 'What kind of filth is that?'

Marc would have laughed, but Fischer didn't need much provocation to smash your brains out.

'It's the ointment they gave me for my eye,' Marc said, trying to keep his voice neutral as Fischer scraped his tongue on a tobacco-stained handkerchief.

'Why didn't you say before I licked it?' Fischer asked. 'Think you're funny, do you?'

Marc hoped the question was rhetorical and didn't answer.

'I'll get one of my men to find you a bed aboard the *Adler*,' Fischer shouted. 'Report back here at eight tomorrow. Make sure you've got something I'll want to hear.'

*

Adler was the largest of the prison boats moored in Frankfurt's East Docks. She had three levels above the hull and four below, and while *Oper* was usually bedded in mud, *Adler* was moored on the riverbank and floated free.

The ship's gentle swaying made Marc queasy as he lay

on the second bunk in a stack of five, two decks beneath the nearest fresh air. There was no ventilation and the build up of cigarette smoke and stench of toilet buckets made it feel like breathing soup.

Marc's mood was black. He'd gone from top to bottom. From friend of the commandant with a cushy job, to number one enemy of a sadistic and unstable guard. From having good mates and an escape plan, to being alone and as far from getting home as he'd ever been.

Adler's prisoners were older than the crowd on the *Oper*. Dutch, Poles and Slavs all mixed together. If any of the fifty bodies packed in Marc's cabin spoke French, he didn't hear them.

The evening meal arrived in a big drum, with black loaves floating in the soup and every man fighting for his share. Marc had left his mug, spoon and mess tin back on the *Oper*, so all he could do once he'd pushed through a greedy mob was grab a chunk of bread and dunk it.

He was no weakling, but had reason to be scared as a new arrival in a cabin full of grown men. Nobody bothered him though, partly because he'd emerged from hospital with nothing worth stealing, but mainly because poor food and heavy work meant the men climbed on bunks and fell asleep, most fully dressed in stinking clothes and boots.

These prisoners were like the living dead: worked to exhaustion, given just enough food to stay alive. Their joyless existence had more in common with that of the cattle Marc looked after before the war than with normal human life.

As Marc lay awake, dripping with sweat and with the fleas in his mattress eating him alive, he realised the prospect of keeping his head down and ending up like his new cabin mates was more dreadful than any threat made by a thug like Fischer.

Escaping now would be almost impossible, but Marc's brain kept cycling back to the same questions: What would Commander Henderson do? How would *he* get out of this?

*

The guards decided it was a good day for a roll call, so at four thirty the next morning five hundred and fifty men who bunked aboard Adler were dragged out of bed to the sound of ringing hand bells, before lining up in rows on the dockside.

The count required every inmate to recite their prisoner number in turn and took forty minutes. When that was done the men were left standing to attention while a small team of guards worked their way through seven prison decks, supposedly searching for contraband, but mainly just throwing mattresses around and occasionally finding something worth stealing.

The guards on the *Adler* were significantly nastier than those on the *Oper*, even when their boss Fischer was nowhere to be seen. As time passed, anyone who scratched, stretched, slouched or moved got slapped with a leather glove if they were lucky, or a rifle butt slammed in the guts if they weren't.

After two and a half hours at attention, the prisoners were sent off to their work details without breakfast. With no work assignment, Marc found a balding guard screaming in his face, calling him a *lazy turd*.

'I have a meeting with Fischer at eight,' Marc said.

'Liar,' the guard shouted. 'Fischer isn't even working today. What's your prisoner number?'

Marc wondered if the guard had made a mistake about Fischer having a day off, or if Fischer's idea of him being his snitch was just an extra way to put the frighteners on him.

He got his answer after several minutes of being yanked around the dockside by his collar, jabbed in the back and shouted at by two different guards until someone found some paperwork with Marc's new work assignment on it.

'Gang sixty-two,' the guard read. 'Get moving.'

The guard who'd been shoving Marc about broke into a high-pitched laugh,

'I don't think Fischer likes you,' he explained, in broken French. 'And such a shame to spoil those nice shoes.'

Before Marc grasped what was being asked, he got smacked up the side of the head.

'Give us your shoes,' the guard shouted. 'How stupid can you bloody French be?'

After handing his shoes to the guard, Marc was dragged over the quayside in socked feet to join up with a dozen wretched-looking prisoners standing under a dock crane. Their clothes were no filthier than any of the construction workers, but the smell of sewage hung over them, even in open air. Worse, many had chunks of missing hair and sores on their skin.

The men began shaking their heads with disgust when they saw Marc. They were mostly Polish, but a couple spoke French, including a red-haired fellow.

'Leonard,' he said, by way of introduction. 'How old are you?'

'Fifteen,' Marc said, figuring it best to stick to his official age.

Leonard translated into Polish and the other men groaned with disgust.

'I can pull my weight,' Marc said defensively.

'It's not that,' Leonard explained. 'We don't like the fact you're so young. Our line of work isn't good for your health.'

CHAPTER NINE

War production put Frankfurt's industry at full stretch. Factories worked 24/7. New facilities opened all the time, staffed by slave labourers living in hastily built camps. This all-out effort led to water shortages and a sewage system on the verge of collapse.

Gang sixty-two weren't trusted to walk the streets in prisoner jackets. They got an armed escort on an uncomfortably brisk three-kilometre walk from the dockside to a Frankfurt Water Company maintenance depot.

Leonard stuck close to Marc as they were assigned a job list and sent out with two other prisoners and a pistol-toting supervisor. Their open-backed cart was packed with shovels, rakes, pipes, hoses and tubs of chemicals.

Marc's first taste of his new job was an open sewer run-off at a women's prison camp. The stench was familiar from every toilet he'd encountered since being taken prisoner, but rolling up trouser legs and wading into a rat-infested lagoon of human waste was all new.

Marc fought dry heaves as he joined the other three prisoners on his team, using rakes and shovels to dig out a soggy blockage made up of newspaper and card that the women had used to wipe themselves.

The next two jobs on the work list were similar. Marc felt sick most of the time and was terrified by the obvious risk of disease, standing barefoot in open sewage. The fourth and final job of the day was a factory, where instead of dealing with sewage they had to clamber into a fume-filled outlet pipe and shovel a build up of toffee-like sludge into wheelbarrows.

There was a disinfectant hose down when they arrived back at the water department at the end of their shift, but it was nowhere near enough to get the stench off clothes and skin.

Marc's second day on the job began well enough when his German supervisor dug out a pair of rubber boots for him, but by afternoon he had a fever and was doubled over with stomach cramps.

Leonard said everyone got sick in this job. He reckoned the first few weeks were worst for picking up infections because you gradually built up immunity. The big long-term danger was exposure to chemicals in the factory run off.

'Losing all my nails,' Leonard said, proving his point by peeling back a yellowed thumbnail that flipped up like a car door. 'A lot of long-termers get problems with their breathing. So far I've been lucky with that.'

That night Marc ran between bed and buckets with diarrhoea. He stumbled off his bunk next morning, shivering and barely able to stand. None of the other prisoners helped because he stank so bad.

'Back to bed,' a guard yelled, when Marc staggered on to the quayside.

Marc felt awful: when it seemed life could get no worse there was always some new depth to plumb. His feverish mind thought about escape, but how could that happen when he could barely move?

The man in the bunk above took pity and fetched Marc some bread and water when the evening meal came. The guard Marc encountered first thing next morning showed less sympathy and forced him to stagger across the quayside to meet up with his gang.

He barely survived the three-kilometre walk to the maintenance depot and his supervisor left Marc behind, hosing down equipment and sweeping the yard.

*

Marc was starting to hope that Fischer had forgotten

about him, but he woke that night with a hand on his throat.

'How's life?' Fischer asked, smiling nastily as his muscular arm drove Marc down into his bed slats. 'Night shift can be boring, you know? Old Fischer needs entertaining.'

A couple of the other prisoners stirred as Fischer dragged Marc from his bed, then marched him ashore to a guard hut by the main exit gate.

The security set up was identical to the *Oper*, with prisoner jackets piled up inside the door and a table where guards took their breaks. But the *Adler* had more than double the number of inmates so there were more guards around.

'Patrol the perimeter,' Fischer told a fat guard, who sat at the table puffing a small cigar. 'I need to have a private conversation with my young friend.'

'I just sat down,' the guard complained, but one whiff of Marc sent him running for the door.

Fischer shut the door with a backwards kick, then shoved Marc hard against the wall, before eyeballing him.

'So, what information have you dug up for Old Fischer?'

'It's hard,' Marc said, trying to hide his fear. 'They're not French. I can't even understand what they're saying.'

'Didn't ask for excuses,' Fischer said, but then stepped back abruptly and laughed. 'Christ, you reek of shit.

Aren't you gonna thank me for setting you up with gang sixty-two?'

Marc scowled, which made Fischer laugh and bunch his fists.

'You want me to wipe that look off?'

Fischer threw a punch, but Marc ducked. This pissed Fischer right off. He pinned Marc to the wall with one knee before launching a volley of slaps and punches that left Marc doubled up, leaning breathlessly against the back wall.

'Listen to the little snitch cry,' Fischer roared, smirking as Marc fought off sobs. 'Tell you what, snitch, how about you get down on your hands and knees? Old Fischer's boots need a good tongue cleaning.'

Marc caught his breath and looked at Fischer's shabby, mud-crusted, boots.

'Crack on,' Fischer ordered. 'I haven't got all night.'

Marc fought pain and tried to remember his training. He felt sure Henderson would be ashamed of him: sick, weak, living day-to-day without any kind of plan. But even though Fischer would hurt him badly, Marc had too much pride to lick Fischer's boots.

'No,' Marc said, shaking his head slowly.

Fischer cupped his ear, as if he couldn't believe what he'd just heard. 'Excuse me?'

'You're not deaf,' Marc said bravely. 'Clean your own damned boots.'

Fischer's punch slammed Marc so hard he felt like his guts would burst. Next Fischer swept Marc's legs away, making him slam the floorboards, hard and face first. Marc expected more blows, but Fischer was surprised by another guard stepping into the hut.

'What?' Fischer shouted irritably, but seemed happier when he saw that it was Osterhagen. 'Recognise Vogel's former pet? The brat who dropped us all in it.'

'What did he do to end up here?' Osterhagen asked, as he recoiled at the stench.

'Night shift gets boring,' Fischer laughed. 'I've got him on gang sixty-two. Sixty hours a week wading through jobbies. That's what happens when you cross Old Fischer.'

Osterhagen had always treated Marc decently and to Fischer's annoyance the young guard didn't seem amused.

'You should be on the *Oper*,' Fischer growled.

'I'd like to request some extra leave, sir,' Osterhagen said. 'My cousin's getting married.'

'You've used your entitlement?'

'Yes, sir,' Osterhagen said, as he pulled a large bottle of cognac from inside his coat. 'My father has a large cellar. I know you're fond of a drop, sir. You're welcome to pick up a few bottles, with my family's regards. Or our butler will happily deliver it.'

'Let's see,' Fischer said, his tone warming as he

inspected the cognac label, then turned to look at a duty roster on the rear wall.

Cognac, butlers and a large cellar confirmed Marc's theory about why Osterhagen wasn't fighting on the Eastern Front with all the other healthy young Germans. But his own situation was of more concern. Two guards with their backs turned was a rare opportunity, but for what?

'If I drop night shift here down from six guards to five, I can send an extra man across to the *Oper* to cover while you're at the wedding,' Fischer said. 'How's that sound?'

Marc glanced about. He hoped to find something he could use to fight back. A loose nail, a bottle, a piece of wood. But the only thing in reach was the mound of grotty prisoner jackets and these sparked another thought: it was dark out and the main gate was only a few metres from the door of the hut.

The gate might be locked. There might be a guard right outside. Marc had no documents, or money. But if Fischer didn't beat him to death, working for gang sixty-two promised a nastier death from chemicals and disease. And when you're already as good as dead, what have you got to lose?

Weak from his illness and the beating, Marc wasn't sure how his body would respond when he tried to move.

'So everyone's happy, except Sivertsen, who's on night duty for the next month and a half,' Fischer laughed, as he gave Osterhagen a slap on the back. 'Would you like to share a quick glass before heading back?'

There was sharp pain in Marc's gut as he stood and his knee buckled as he crept towards the door, which Osterhagen had mercifully kept open. He slid a prisoner jacket off the pile and crawled out, as Fischer and Osterhagen clanked their glass tumblers.

'Cheers,' Fischer said.

'Good health,' Osterhagen replied, as Fischer saw Marc reflected in his glass.

'You dare move!' Fischer roared, making Osterhagen jump.

Marc limped out of the hut and turned towards the gate. The only guard in view was the fat man who'd been in the hut. He sat twenty metres away, finishing his cigar on a bollard at the water's edge. Marc went for the gate, but found it padlocked. The chain-link fence was climbable, but it was topped with barbed wire and Marc was in no state to climb quickly.

Osterhagen was first out of the hut, with his baton drawn. Marc glanced across the open quayside where they lined up for roll call, but the whole area was fenced in and even in moonlight he'd make a nice easy shot for any guard with a rifle.

A trip up the *Adler*'s gangplank was Marc's only option, but he had no idea what he'd do once he got there.

Osterhagen was surprised to see Marc running towards him and only managed a clumsy baton swipe as he swept past. The guard on the bollard got to his feet. He would have easily intercepted Marc at the base of the gangplank, but he froze stiff when Fischer's pistol went off.

Marc expected a bullet in the back as he ran up the gangplank, but Fischer was hopeless, his bullets splintering the wooden handrail and clanking against the hull several metres off target.

'You're better off dead,' Fischer shouted, furious at his poor marksmanship. 'I'll shatter every bone in your body.'

At the top of the gangplank Marc faced double doors in one direction and a section of warped decking leading up to *Adler*'s bow in the other. He crouched low, but before he got a second to think a fourth guard who'd heard the shots charged through the doors.

'What's going on?' he shouted, unaware that Marc was so close.

Marc grabbed a life-preserver ring and used everything he could muster to whack the guard in the face with it. As the guard crumpled, clutching a bloody nose, Fischer was charging up the gangplank with his pistol poised.

Marc jumped on top of his victim. His first thought was to snatch the rifle hung over the dazed guard's shoulder, but the man's jacket was open, exposing a ceremonial dagger with a red and black swastika embossed in the handle.

'Hands up,' Fischer shouted, as he came around the top of the gangplank, aiming the pistol from less than three metres.

Marc had decided to die rather than surrender. With a single movement, he rolled off the guard and threw the knife. Marc was an expert knife thrower so his aim was no accident, but you need to gauge the weight of a knife and throw it a few dozen times before you get any sense of how it flies, so it was more by luck than judgement that the blade speared Fischer's heart pointy end first.

Fischer squeezed the trigger as he staggered backwards towards the bow. Marc sniffed gunpowder and felt a thud. He thought he'd been hit until he saw the shattered hip of the guard lying on the ground beside him.

Blood poured across the deck as Marc freed the rifle.

Fischer was thrashing about, trying to rip the knife out of his heart as blood foamed out of his mouth.

Marc worried that another guard could burst out of the doors behind him, but there was enough light escaping the guard hut for him to see Osterhagen sprinting towards the telephone by the main gate. Marc took two shots with the rifle. The first only tore a lump out of the quayside concrete, but the second smashed Osterhagen in the base of the spine.

Marc ripped the pistol out of Fischer's hand, guessing that there could only be one or two shots left. The fat guard was still around somewhere and Fischer had just told Osterhagen that there were usually six on duty, so Marc kept low as he crept down the gangplank.

When Marc reached the bottom, fatty was squatting with his hands raised in surrender. He looked harmless, but Marc didn't have time to faff about disarming him or tying him up.

What would Charles Henderson do?

'Please,' the German begged, backing up as Marc swung around with the rifle and squeezed the trigger.

The muzzle lit up and the fat man's head clanked against the *Adler*'s hull before splashdown in the river. Marc did a 360, looking for more guards, then slung the rifle over his shoulder as he limped quickly towards the gate.

Osterhagen was a mess, face down with a bloody hole in the back of his jacket and passed out in shock. Marc felt guilty as he crouched down and ripped the blood-soaked keys off his belt.

The telephone receiver was dangling and Marc heard a crackly voice on the other end going, *Did we hear shots? We're sending a response team over*, as he tried the keys in the lock. Marc considered telling them things were fine, but while he spoke decent German he reckoned his age and

French accent counted against passing himself off as one of the guards.

When Marc got to the fifth key he remembered that Osterhagen worked on the *Oper*, so why would he have the key for this gate?

Marc decided to look in the hut. If that failed he'd have to go back up the gangplank to get them off Fischer. When he turned back towards the *Adler*, Marc noticed a few prisoners out on the deck, trying to see what had happened.

To Marc's relief, the gate key was inside the hut, attached to a huge wooden key fob with *Main Gate Do Not Remove* written on it. As Marc walked back to the gate, a few curious prisoners had ventured as far as the gangplank.

They seemed happy to find Fischer dead, but there was no prospect of a mass breakout. These were the exhausted, timid men that Marc was terrified of turning into. Prisoners were already heading for refuge in their bunks before reinforcements arrived.

As the padlock popped and the gate swung, Marc noticed a silver bike resting against the fence. It was a racing bike, the kind only a rich, young gent like Osterhagen could afford to own. With a rifle over his shoulder and a pistol inside his jacket, Marc straddled the bicycle and pushed off.

Adrenaline had got him this far, but Fischer had

battered him and he was running on flat batteries as he pedalled away, with weak legs and fingers barely able to grip the handlebars.

CHAPTER TEN

Marc's first priority was to put distance between himself and the scene of his escape. He raced along the riverbank, heading towards Großmarkthalle for no reason beyond the fact that Osterhagen's bike had been pointing that way.

The hugeness of what had happened sat in Marc's gut like a rock: deep in Germany, no money, passes, or maps. If they caught him, his death would be slow and probably in front of several hundred prisoners to make an example of him.

Marc had no watch, but he guessed 3 a.m. There was nobody on the waterfront and no artificial light because of the air raid threat. This gave Marc some anonymity, but a Hitler Youth or police patrol might lurk around the next dark corner.

Many prisoners worked nights, and Marc's prisoner jacket would get him past all but the most persistent of patrols. But no prisoner ever rode a bike – especially a beauty like this one.

Once he'd ridden a kilometre and a half, Marc rolled up to the dockside wall, took a glance around and felt sad ditching the best bike he'd ever ridden in the river. He also ditched the rifle because it looked obvious, but he kept Fischer's pistol tucked inside his trousers.

Now what?

Marc reckoned he had ten minutes – twenty at best – before the alarm went up and there were teams of Germans hunting a killer. On the upside, with Fischer dead, Osterhagen unconscious and the other two guards never having known him, it would probably be much longer before the Germans knew exactly who they were looking for.

So Marc had a window. Central Station was only a fifteen-minute walk, but Marc doubted any passenger trains ran at this time of night. Even if they did, he had no ticket money and there was heavy security at major stations even when someone hadn't just killed a bunch of prison guards.

If Marc had been fit, running into hiding somewhere outside town would have been possible, or he might even have swum across the river. But he could barely walk, so his only realistic route out of town was sneaking aboard a truck or cargo train.

Großmarkthalle was less than three hundred metres from where Marc had ditched the bike, with goods sidings branching off behind. Only a dozen or so trains left the sidings per day, so he'd need luck to find one departing any time soon.

A cart or truck seemed a better bet. Local factories worked through the night and the market hall stayed open to accept their deliveries. Marc had never been inside the hall much after 10 p.m., but he knew that a shift of prison labourers came on duty at around 6 p.m. and worked through the night.

Marc's first thought was to walk to the rear of Großmarkthalle and sneak in through the railway sidings. But he'd entered the building hundreds of times with no hassle from the guard on the entrance, so being sneaky was probably the riskier option.

Fischer's kidney punch meant Marc clutched his sides as he walked, but he took a deep breath, straightened up and tried disguising his limp as he approached the gate.

'Haven't seen you in a while, son,' the guard on the gate noted.

It would have been better not to have been remembered, but Marc was happy enough as he trotted down eight concrete stairs and along a short corridor before passing into the main hall.

To save electricity the cavernous space was barely lit.

The high ceiling and shafts of light had an eerie stillness, as did the endless pallets, drums and boxes stretching over several football pitches worth of concrete floor.

'Haven't you heard of a shower?' a pot-bellied Frenchman grumbled, as he buried his face behind his arm.

He was having a sneaky cigarette break with two fellow prisoners, who pinched their noses shut like five-year-olds smelling a fart. Marc couldn't smell himself, but it was an unwelcome reminder that the filth soaked into his clothes made him stand out.

He cut between the piles of goods, moving as purposefully as his physical limitations allowed. If tonight was anything to go by, prisoners working the nightshift at Großmarkthalle had an easier job than most. About a dozen men were rolling huge drums of copper wire on to a goods wagon, but twice that many seemed to be sitting about.

There was no sign of a train steaming up ready for departure. Even more disappointingly, all but one of the doors where trucks and carts backed in to deliver goods were shuttered, so his chances of sneaking out that way were nil.

Nobody had paid much attention so far, but it was only a matter of time before someone asked Marc what he was up to. He did his best to look purposeful and had almost walked the entire length of the market when noise broke out on the metal stairs leading up to the offices.

It was near dark, but Marc still dived behind a stack of boxes, before peeking up to see a dozen agitated Gestapo men bolting down from their fourth-floor offices.

Clanking treads drowned most of what they said, but they were clearly men in a hurry and Marc caught the German words for *perimeter* and *guard*, so he was sure they'd been sent out looking for him.

The streets would be much more dangerous now, and with four guards either dead or seriously wounded Marc knew they'd use round-up tactics, where everyone on the streets would be scooped up on sight and brought in for detailed interrogation.

But it was no less dangerous hanging about on the market-hall floor and if Marc couldn't leave, the only choice was to hide. His first thought was to sneak under a pile of sacks, or make a crawl space between boxes, but it was a short-term solution at best, and his smell would work against him.

Then Marc considered the fifth- and sixth-floor offices where he used to work. He couldn't decide if it was genius or the worst idea ever. But his kidney was burning, his knees were getting weak and alternative plans weren't exactly forming a queue.

After giving the Gestapo team half-a-minute to clear out, Marc cut around the edge of the empty Gestapo prisoner pen and started up the stairs. He was on the fifth flight of twelve when a trio of Gestapo men started belting down the stairs, one with a machine gun slung over his shoulder.

'It's Baron Von Osterhagen's grandson,' one of them said, as they closed on Marc. 'He owns half of Frankfurt. Heads are going to roll.'

'Serves the rich little prick right for draft dodging,' another officer said harshly.

'You might be right, but I wouldn't voice that opinion too loudly. Nobody messes with the Baron in this town.'

Marc paused on the next landing, trying not to shudder as the three Gestapo men swept past, giving him looks of utter condescension. He half expected to hear a gun click, or someone to turn back and say, *hang on a minute*, but Marc's undernourished teenage frame clearly didn't fit the officer's mental image of someone who'd just taken down four guards in a daring escape.

Marc walked to the sixth floor, because the archives were always quieter than the busy Labour Administration office downstairs and had the additional benefit of access to the roof. It was no surprise to find the door locked. Even if he'd had the strength, smashing the lock would give Marc's presence away, so he had to pick it.

He quickly tapped down his pockets, but found no obvious pick and realised he'd have to strip the pistol.

There wasn't much light up this high and Marc wasn't familiar with the design of Fischer's automatic. But all automatic pistols have a recoil spring inside, and after half a minute's fiddling, Marc got the weapon's outer casing off and freed the elusive spring.

Marc's grip was weak and he felt nauseous as he used all his strength to stretch one end of the spring into a length of thickish wire, with a slight L bent into one end. He crouched on one knee, turned the lock handle with one hand, then pushed his freshly made lock pick into the hole.

Mercifully it was a simple lock. It only took a few jiggles, followed by a sharp hammer on the end of the pick with the pistol. The door opened inwards, taking Marc with it. After the undignified entrance, he nudged the door shut with his rubber boot, then spent several seconds sprawled on the floor, clutching his chest, while close to blacking out from pain and exhaustion.

When Marc started to move it was a half stoop, half crawl towards a tiny windowless washroom. He flipped on the light, let the gun parts clatter in to the sink and made a long dry heave over the toilet. His kidney was on fire. Too weak to keep standing, he dropped trousers, sat on the toilet and started pissing something the colour of red wine.

'Bastard,' Marc muttered to himself, as he thought about Fischer.

But it was good knowing Fischer was dead and he even raised half a smile at the thought that if he got caught and executed, he'd taken a few Germans with him.

After a long drink of water from the cold tap, Marc studied himself in the mirror. He'd not seen a mirror since he'd left hospital. His eye was the only bright spot: the surgeon had done a good clean-up job, and despite all the sewage he'd been exposed to there was no sign of infection, just a little swelling and an X-shaped scab above his eye.

The other news wasn't so good. Marc's shirt and jacket were crusted with dried sewage and his skin was filthy and wrinkled like an old man's. Unbuttoning his shirt revealed a mass of bruises and semi-circle cuts made by the ring Fischer wore on his right hand.

Marc knew it would be a couple of hours before the caretaker arrived to unlock the office. It gave him a breathing space, but he had no idea how to pull off the next stage of his escape.

CHAPTER ELEVEN

Marc put the pistol back together. There were three bullets in the clip, but with no recoil spring he'd have to load each shot manually.

With time to think and get his strength back, he was able to put his espionage training to proper use. Escape routes were a priority, so he found an emergency bolt hole behind two cabinets, then opened the door out on to the balcony.

The cool, still air sent a shiver down his back as he turned around. His fingertips could reach the edge of the flat roof, but he wasn't currently strong enough to pull himself up, so he put a wooden chair outside to make the climb easier.

With hiding and escape plans in place, Marc's thoughts turned to documents. He didn't dare turn on the electric light in the main office because it would be seen from below, but he found and lit one of the paraffin hand lamps that office workers used during Frankfurt's all-too-frequent power cuts.

It seemed Commandant Eiffel had ordered no new security precautions beyond Marc's sacking. Valuable documents such as blank prisoner record cards and travel permits were still piled up in an unlocked cupboard, just like when he'd worked here.

Marc grabbed a couple of each, then headed towards the cabinets to retrieve his own prisoner record card, hoping he'd be able to use the photo on a travel warrant. But his replacement card was only a few days old. His new photograph was still being processed and the only things written on it apart from Marc Hortefeux and a false date of birth was a message in block capitals stating that ALL MATTERS RELATING TO THIS PRISONER MUST BE REFERRED TO COMMANDANT EIFFEL.

'Shit,' Marc said, deterred, but not beaten.

The cabinets contained thousands of cards. Pictures were usually taken when prisoners were first arrested. The resulting photographs depicted men and women who were noticeably plumper and hairier than they'd be after a few months' captivity.

Marc flipped through rapidly. Within minutes he'd picked out the photos of two teenaged prisoners who looked quite like he'd done before he'd lost twenty pounds and had his head shaved.

The next job was to type up the false prisoner record and travel warrant, then he'd have to sneak down to the fifth floor where he'd find the rubber stamps that validated documents and a fountain pen to forge signatures.

But before Marc reached the typewriter, he heard a key in the lock. After quickly snuffing out his lamp, he bolted towards the balcony at the rear of the office, making it out of the door as someone flipped on all the lights.

Marc had the chair ready, so that he could get up on to the roof if he needed to, but for now he crouched on the balcony, peeking down the rows of filing cabinets towards three figures who'd entered the archive room.

Two were Gestapo men. One plain clothes, one the local Gestapo boss, decked out in his black uniform. The third person was Commandant Eiffel, but the way the men stood slightly behind on either side gave Marc the impression that she was in trouble.

Eiffel led the way to the exact drawer that Marc had been in moments earlier. He could see wisps of paraffin smoke where he'd hastily extinguished the lamp, and hoped they weren't noticed.

'Is your toilet blocked?' the Gestapo chief asked, as he sniffed suspiciously. 'It smells vile up here.'

'I've never noticed before,' Eiffel said, as she pulled the draw open. 'But it's quite a smell. I'll get the caretaker to take a look at the plumbing . . . Oh, damn.'

Eiffel had pulled Marc's record card and discovered that his photograph hadn't been processed.

'This is unacceptable,' the Gestapo chief yelled, as he snatched the card and read it. 'How can your photographic records not be up to date?'

'There's a shortage of photographic paper,' Eiffel explained. 'A delivery is expected within the next week, but Reich Labour Administration stands well below the military and many other departments when it comes to items in short supply.'

The Gestapo commander turned towards his plain-clothes colleague. 'I want *everyone* who's had contact with the escaped prisoner rounded up for interrogation. People he works with now, people he worked with in this office, guards, people he met while he was in hospital, his former cabin mates on the *Oper*. One of them will know where he's heading.'

Eiffel slouched like a little kid about to get yelled at by her dad as she led the Gestapo officers towards the door and switched the lights back out. Marc felt bad, knowing that old mates like Richard, and Vincent, along with friendly co-workers like Ursula and Leonard now faced interrogation, possibly even torture because of what he'd done.

Marc's biggest worry was someone working out why the sixth-floor office stank when they met his co-workers on gang sixty-two, but there were no risk-free choices and hiding out here remained a better option than trying to skip town through heavily policed streets.

Having decided to stay here for at least a few hours, Marc prioritised checking the roof for hiding spots and escape routes over typing up documentation. Using the chair as a step, Marc clambered on to the flat roof of the offices, which overlooked the huge vaulted roof of the main hall a few metres below.

The only comfortable hiding place was an open-sided wooden shelter, built next to the turntable for an anti-aircraft gun. Frankfurt was at the outermost edge of British bombing range, so although Großmarkthalle was a high-value target, the gun had been removed to another location where air raids were more frequent.

The shed was such an obvious hiding place that Marc scouted the edge of the roof looking for a plan B. All he found were antennas and a couple of precarious ledges where roof gutters met rainwater pipes. He reckoned he could climb on to one of these and crouch down out of sight for a few minutes, but one slip would send him plummeting six storeys, so it was strictly a last resort.

Marc's next two priorities were washing his clothes and finding food. German civilian rations weren't much better than what prisoners got. Marc knew he wouldn't find much food lying around the office and the matter of filling his belly would require further thought.

Washing his clothes was simpler. There was no soap, but Marc found a bucket in the cleaners' cupboard along with a pack of chlorine powder used for scrubbing the office floors. He didn't want to get trapped in the bathroom if someone came in, so he filled the bucket and waddled to the balcony, careful not to splosh water on the floor.

It took half an hour and two fresh buckets of water, starting at the top with his hair, scrubbing his body, then he washed his shirt, trousers and socks before finishing off by wiping down his rubber boots. The chlorine powder was the same stuff used in swimming pools, but Marc used it at a higher concentration. He had to avoid breathing the fumes, which reddened his eyes and burned the small cuts on his hands and torso. But after the demoralising effect of being covered in raw sewage, the stinging cleanliness was almost therapeutic.

The cleaning cupboard had spare towels for the commandant's bathroom, and since prisoners weren't issued towels, this was the first time Marc had towelled off since the day he'd been arrested. He wrung out his clothes, but couldn't let them drip all over the floor so he hopped back inside naked.

He dried the inside of the bucket so that the cleaners didn't suspect anything, then carefully replaced everything he'd used except for the damp towel, which he thought might make a good pillow. As he walked back to the balcony, Marc peeked over the filing cabinets and glanced down through windows that overlooked the inside of the market hall.

The lighting had been turned up and the wooden pen where the Gestapo usually kept prisoners awaiting deportation was filled with the fifty-odd men with whom Marc had shared a cabin on the *Adler*. They stood to attention in five short rows, with a Gestapo officer shouting so fast that a prisoner enlisted to translate into Polish couldn't keep up.

'I want you to think carefully,' the Gestapo officer shouted. 'Telling me what you know is in your interest. If Hortefeux is not found, I shall draw lots and every third man will be sent to punishment camp.'

Marc couldn't hear all that was said, but the situation was clearly farcical. Most of these Polish and Flemish speaking prisoners couldn't have held a conversation with Marc if they'd wanted to. The Gestapo officer changed tack and asked who spoke French, but nobody was prepared to admit a skill that would apparently lead to further interrogation.

At the same time, Marc saw the arrival of all the secretaries who worked downstairs. The two old battleaxes came fully dressed, but the younger women looked very different without their hair and makeup done. Some were even barefoot or in nightclothes, where the Gestapo had dragged them out of bed for maximum intimidation.

Marc became engrossed and got caught out when the door opened again. This time a pair of middle-aged police officers came in. They were lumbering creatures, clearly unhappy to have been dragged out of bed to join a city-wide search.

Marc scrambled away from the window and ran naked to the balcony. He scooped up his boots and wet clothes, then threw them up on to the roof before stepping on to the chair and climbing up after them.

A police siren down in the street made Marc even tenser as he lay flat on the roof, looking down at the balcony with grit sticking into his bare skin.

He waited more than ten minutes and was starting to think that the cops had left when the door out on to the balcony finally opened. As the cop shone his torch, Marc saw that the balcony floor had several damp footprints where he'd washed and one of his soggy socks balled up close to the chair leg.

'Nice view up here,' one officer noted, switching his torch off as he stepped out on to the balcony. 'That's the church where I got married, across the river.'

The other officer was younger, but still bald and the wrong side of forty. 'Didn't know you were married, boss.'

'My marriage is too traumatic,' the older officer laughed, as he pulled a pack of cigarettes from his pocket and offered one to his colleague.

Marc's heart pounded and he slid the pistol out of the prisoner jacket just in case. The younger man was looking around, but fortunately with the torch extinguished it was much harder to see the sock or the damp footprints.

'You could get up on to the roof easily enough from here,' the younger cop noted.

Marc pulled his clothes bundle tighter and got ready to run as the older officer looked around. The gun wouldn't reload without its recoil spring, but with luck he could shoot one officer with the bullet in the chamber, jump down on to the market hall roof and run along the metal gantry until . . .

Until what?

'You're *more* than welcome to go clambering over rooftops,' the older officer laughed, as he lit a match for the cigarettes. 'But I won't be joining you. They say this kid is fifteen. Is he really gonna run to Gestapo headquarters to hide?'

'You mark my words. He'll be squatting in a garden, or lying under a car. He'll move when he gets hungry and they'll pick him up at the first checkpoint he comes to. If it wasn't the Baron's grandson, I'd probably still be in bed listening to my wife snoring her ugly head off.'

'The Standartenfuhrer said he smelled sewage up here,' the younger officer said.

'That pompous arse either smelled his own bullshit, or that stuck-up bitch Eiffel chuffed one out of her drawers.'

'I still think . . .'

The older cop interrupted, sounding irritated. 'We'd get bird shit all over our uniforms. Stop going on about it. Relax, smoke and watch the world go by. You'd better learn to take things easy if you're gonna live to be an old man like me.'

CHAPTER TWELVE

Marc huddled in the rooftop shelter, damp clothes spread out to dry, towel under his head and the grubby prisoner jacket as a blanket. If they found him up here there was no escape, so he kept the gun close: better to blow his own brains out than go through brutal interrogation and die a few days later.

Hunger and nerves made sleep hard, but Marc clocked up a few ten- or twenty-minute bursts. It was one of the warmest days of the year so far and thirst began to torment him as morning broke.

Occasionally he heard muffled conversations, or a cabinet drawer slamming in the archives below. A few times he thought he heard footsteps on the flat roof, but pigeons were the only company that existed outside his imagination.

The day passed agonisingly. Marc started feeling light-headed as his body demanded food and water, but it would be insanely risky to venture down into the office before the Labour Administration staff clocked off at 6 p.m. In some ways the hunger was a relief: having an appetite made him hopeful that he'd fought off the stomach bug.

With no watch it was hard to judge time, but when Marc finally saw the sun dropping behind smoking factory chimneys, he decided it was nearer to seven than six and made his move.

His clothes were dry, if a little stiff. The chlorine had partly bleached his trousers, leaving them colour streaked and a few shades lighter. This was actually a lucky accident, because the Gestapo would have picked up the matching jacket that he'd left on his bunk when Fischer dragged him out of bed.

The chair had been taken back inside, so Marc had to swing over the roof's edge and drop on to the balcony. It would have been no problem when he was fit, but the landing buckled his weak knee and sent waves of pain through his chest and legs.

The balcony door didn't have a lock. Marc peered cautiously into the sixth-floor office and was pleased to see all lights off and nobody home. He trembled as he dashed the length of the office towards the toilet. The wall clock said 6:25, which was earlier than he'd have liked, but his only concern was sticking his head under the cold tap and gulping his first drink in fifteen hours.

Food was Marc's next thought. He ventured down to the fifth-floor offices after ten minutes carefully listening for signs of life. The open staircase gave a clear view over the market hall. There was a rush down below, with prisoners frantically loading a pair of trains in the sidings out back.

Pallets, barrels, tank tracks, bags of horse feed and even huge artillery pieces had to be lifted on to open carriages by two-dozen prisoners using a manual hoist. German supervisors motivated their staff with angry shouts and riding crops.

It was the kind of chaos Marc had hoped to run into the night before, but he now had no intention of leaving without money and a much clearer escape plan.

He was pleased to find the fifth-floor office locked, which meant there was no late meeting, or last-minute letters being typed inside. It was the same type of lock that he'd popped on the sixth floor the night before, and the benefit of experience meant he cracked it with the bent recoil spring in seconds.

Thinking of his stomach, Marc opened the secretaries' desk drawers, hoping to find something. When that failed he rifled through bins. His quest for a crust or apple core was fruitless, but he scored a minor haul in Commandant Eiffel's office.

She must have had an important meeting earlier in the day, because a tray with plates and used coffee cups rested on her sideboard. It would have been Marc's job to clean this up when Vogel was commandant, but Eiffel had left it for the cleaners.

Sugar was in short supply, but a couple of spoonfuls remained in the bottom of a silver bowl, along with the dry edge of a pastry and a blob of jam stuck to the side of a plate. Marc dived in, tipping the sugar granules into his mouth, cramming down the pastry, dabbing up all the loose pastry crumbs and washing it all down with a splash of cold and revolting ersatz coffee[5].

It was less than a tenth of the daily calories a teenager like Marc needed to survive, but the sugar rush gave an instant energy boost. As his mind came into focus, Marc considered his main dilemma:

Over time, the Gestapo would start to believe he'd already left the area and scale down their search. If he could wait another day or two, he'd be able to skip town more easily. But he was already feeling light-headed. There was no food here and if he didn't eat properly soon, he wouldn't have the strength to escape.

*

Marc spent the next hour working up a set of documents.

---

[5] Ersatz coffee – the Germans had no access to coffee-growing areas and made this bitter-tasting fake coffee from acorns.

Travel permits were only valid for a journey commencing on a specific day, but with an unlimited supply of blanks he typed himself permits for each of the next four days. He gave himself different names and destinations, then typed up prisoner release letters that he could use to get identity and ration cards from an administrative office when he arrived in France.

When everything was stamped, signed and triple-checked for errors, he put the whole lot in an envelope, along with the two lookalike photographs he'd found the night before. He couldn't stick the pictures down until he knew which set he'd end up using.

The RLA regularly sent prisoners from one place to another, so the office had a railway map and timetables for most of the major routes.

Marc had already studied German trains when he'd arranged to escape with his cabin mates aboard the *Oper*. The Frankfurt–Bonn–Paris route he'd selected then was fast and direct, but that plan had relied on having nobody missing them until after they'd reached France.

Marc was now the most wanted fugitive in Frankfurt, so he needed to do something more sophisticated than simply going to the main station and booking tickets for the most obvious route home.

He spent thirty minutes studying alternatives, including travelling from local stations where security was likely to be more relaxed, and the possibility of throwing the Gestapo off the scent by travelling deeper into Germany before boarding a train that crossed into France.

He wrote down several routes and lists of train times, but no one route stood out. They all had their own set of risks and danger, and the longer he studied the railway map the more his thoughts turned back to his growling stomach.

Marc had thought up two options for food. The first was to sneak down into the market hall, locate crates of canned food waiting to be loaded on to a train and steal some. But food pallets were always well guarded from theft by the hungry prisoners, there would be an investigation when the damaged boxes were found and his diet would be restricted to a single type of food.

The second option was to use the staff canteen. Marc had collected Commandant Vogel's lunch on many occasions. Although the Reich Labour Administration was a civilian organisation, the commandant and senior staff who worked in Großmarkthalle were given a military ration ticket, allowing them to eat in the canteen.

Vogel always kept his lunch tickets in his desk, but Marc had already searched the commandant's office when he'd been looking for food and apparently Eiffel carried her card around. But Marc knew that unissued ration cards were kept in a locked metal cabinet.

He could have picked the lock, but didn't even need

to do that because Vogel was a forgetful soul who kept spare keys taped behind the toilet cistern in his private bathroom. If he'd told Eiffel about them before he got sent east, she hadn't bothered to move them.

Lunch ticket books were stamped with the owner's name when they were issued, but individual perforated tickets could be torn off and used separately, allowing staff to pick up food for colleagues.

Marc swiped a brand-new book with fourteen tickets, along with a few coins from the petty cash tin, from which Vogel or one of the battleaxe sisters used to give him money when he was sent to the post office with telegrams. Taking a larger sum would be noticed, so Marc decided not to risk that until he was leaving for good.

Wearing his prisoner jacket and trying his best to look confident and purposeful, Marc moved downstairs, shuddering involuntarily as he passed the fourth-floor landing in front of Gestapo headquarters.

One of the trains was still being prepared for departure. He dodged prisoners pushing overloaded carts and made it down the hallway to a basement canteen, earning nothing more stressful than a curious glance from one of the German supervisors.

Marc assumed the greatest risk would be getting recognised by one of the canteen staff, and he kept his hand on the gun in his pocket as he exited the main hall and walked down ten metres of corridor.

Hunger made even the sulphurous smell of stewed greens appealing. What Marc hadn't anticipated was stepping in and finding the serving counter shuttered and nothing in the seating area but condiments, crumbs and a few scrunched newspapers.

He glanced at a couple of the papers. One was a Frankfurt evening paper and had the bottom third of the front page dedicated to his escape. The headline: *Hunt for boy prisoner. Heir to Von Osterhagen fortune among dead.*

Osterhagen was a decent guard and his death saddened Marc, but he was pleased to find no picture of himself, just a vague description: young, 170 cm, fair hair, speaks good German.

The counter was shuttered from the inside, but the door alongside it had the same simple lock as the ones he'd already picked to get into the offices. He listened at the door for a few seconds to make sure no one was behind it, then used the pistol spring and a good bang with his palm to open up.

The lights were off inside. There were big ovens and gas hobs, and it was sweltering where they'd been running for most of the day.

Marc was drawn to the larder at the back of the room. It would have been safest to grab as much as he could and leave quickly, but he was so hungry that he couldn't resist grabbing a cook's knife and slicing a chunk from a hanging bierschinken sausage.

Marc's taste buds erupted as his mouth filled with the mixture of pistachio nuts, ham and garlic. Next he went for a chunk of rich white sausage, made from veal, cream and eggs. The food he'd picked up here for Commandant Vogel had always been plentiful but basic, so Marc guessed this luxury fare was reserved for senior Nazi officials.

After a few indulgent mouthfuls, Marc realised he needed to be sensible. He'd only eaten one small meal since his bout of illness and it was better to eat plain food than to gorge on rich stuff and end up spewing.

He'd only have been able to get a couple of meals if the canteen had been open, so the closure was a bonus. Marc found a small cloth sack with a few grains of barley in the bottom and began stuffing it.

He wanted high-energy foods that would stay edible for a few days, so he went for cheese, sausage, canned pork, a large tin of condensed milk, sugar, biscuits, tinned fruit and a jar of chopped nuts. Finally he went back into the kitchen and grabbed a spoon, a tin opener and a couple of cook's knives which he thought would be good for throwing.

'Anybody home?' a German shouted.

Marc jolted, spun around and ducked behind a metal preparation surface. He could have sworn he'd shut the kitchen door on the way in, but apparently he'd not fully pushed it up. It had drifted open and the bald head of a

German supervisor now poked through it.

Marc was furious with himself and scared by how easy it is to make mistakes when you're weak with hunger. As the German stepped in, he crept backwards into a tight space between a bunch of flour sacks.

'Hello, hello?' the supervisor said, sounding more guilty than suspicious. 'Somebody left the door unlocked.'

Marc pulled his legs tight to his chest and moved one of the sacks so that you'd have to walk right up to the back of the kitchen to spot him. But the supervisor just assumed the door had been left unlocked by mistake, and only seemed interested in the contents of the larder.

Marc heard but didn't see as the supervisor picked up the knife he'd himself used to cut sausage. This was followed by chewing sounds and a big, *mmm*.

With the cloth sack in one hand and the knife in the other, Marc crawled out of his hiding spot as he heard the supervisor moving deeper into the larder. As the delighted supervisor pocketed a can of pear halves, Marc peeked around the kitchen door to make sure it was all clear before dashing out and heading along the hallway towards the stairs.

CHAPTER THIRTEEN

Food and drink brought Marc's strength up and enabled his brain to focus on something other than a growling stomach. Better still, after months of powerlessness, being master of his own destiny felt good.

He didn't want to risk spending any longer in the offices than he had to, so he went straight back out to the roof with his booty. He ate and drank slowly and was greatly relieved that the food seemed to be staying down. Then he got a feel for the two cook's knives by repeatedly throwing them into the side of the wooden shelter.

A maintenance crew cleaning skylights in the market hall roof took the wind out of his sails, but they disappeared after an hour and Marc watched the sunset, lying on his back while scoffing tinned peaches.

He slept well – perhaps too well for someone in so

much danger. When he woke it was light. There were already staff working in the offices below, though it was a Saturday so they'd only work until lunchtime.

Marc suffered fewer aches and pains than the day before, though his ribs were badly bruised. Breathing deep was excruciating from where Fischer had punched him in the kidney, but the damage was healing, at least judging by the clear urine when he peed in the empty peach tin.

He decided to make his move today: half-day in the office gave him an opportunity to leave while the streets were busy and catch an afternoon train, and as very few passenger trains ran on a Sunday he'd have to risk two more nights hiding out on the roof if he didn't.

When the staff left, Marc picked up everything from the roof and headed inside. He dumped his litter behind one of the file cabinets, where it wouldn't be discovered until he was either home safe or dead. After a quick wash in the bathroom, he stole another of the commandant's clean towels and headed down to the fifth.

Commandant Eiffel had worked an extra hour and was just locking the door. Mercifully she didn't look up as she waited for the lift. Once she was out of the way, Marc picked the lock then completed his final tasks: sticking the photos to the travel warrant, adding today's date and stealing a larger sum from the petty cash tin.

As he headed out, Marc remembered that the guard

on Großmarkthalle's exit occasionally asked why he was leaving, so he grabbed a couple of Deutsche Post's yellow telegram forms, folded them in three and kept them in his hand.

It was heart-in-mouth time on the stairs as three Gestapo officers came by, but their minds were focused on a stunningly beautiful teenager being dragged up by her hair. She had a Star of David on her dress and when Marc neared the ground floor he saw that the big wooden pen was crammed to bursting with more than two hundred Jewish women.

To reach the exit Marc had to walk alongside the pen under the gaze of half a dozen SS men guarding the Jews. A frail palm with a folded note inside shot into Marc's path.

'Post this for me,' the woman begged.

Her voice was weak. Marc made an instant decision and snatched the letter. At the same moment, he could hear women arguing on the other side of the pen. Their German was faster than he could follow easily, but the gist of it was that they believed that the pretty girl had been dragged upstairs to be raped rather than interrogated.

'Do you like the Jews?' a guard asked, stepping in front of Marc with a big Alsatian at his side. He had a few days' growth of beard and his uniform was dirtier than any German Marc had seen until now.

Marc acted dumb. 'No speak German.'

'Give,' the guard said, before snatching the letter. 'You wait. My colleague speaks French.'

As the guard beckoned a colleague with his gloved hand, Marc frantically waggled the yellow telegram papers and made running motions with his arms.

'Standartenfuhrer, urgent!' he said.

Marc's experience with Germans had taught him that their harsh regime made everyone afraid of upsetting their bosses.

'Urgent,' he repeated. 'Telegrams for the Standartenfuhrer.'

This did the trick. When the French-speaking officer arrived, he glanced at the folded blank telegram forms and pointed Marc towards the open doors where the trucks delivered cargo.

'Go that way, it's faster. But *don't* interfere with the Jews again.'

Marc nodded and turned back on himself. When he re-passed the woman who'd handed him the note she looked upset, but gave him a nod of thanks for trying. He was almost at the door as he heard a wail. The French-speaking guard had dragged the elderly woman over the edge of the pen and punched her in the side of the head.

'Stay back from the edge,' he roared, as the old woman collapsed in sobs.

Marc felt disgusted as he jumped off the ledge where

the trucks backed in and started walking up a ramp. Pretty girls raped, Jews beaten and herded like cattle, bastards like Fischer. It was like the devil himself had been put in charge.

*

While horrors unfolded behind brick walls and barbed wire, Frankfurt didn't seem such a bad place on this sunny Saturday afternoon. Marc strolled past kids playing football in a park, hanging flower baskets and smart trams rattling down cobbled streets.

The city had only been pricked by minor bombing raids and you'd hardly have known there was a war on, but for the shades over car headlamps and the noticeable lack of young men.

Central Station was a twenty-minute walk if you went direct, but Marc's paperwork said he was a French civilian worker heading home on compassionate grounds, so he needed to ditch his prisoner jacket.

He'd eyed a dilapidated riverside residential district when he was up on Großmarkthalle's roof. The reality of it was grim, with six-storey apartment blocks built along alleyways narrow enough for a man's fingertips to touch the buildings on both sides. Washing zig-zagged overhead. Not much light reached ground level and the smell of drains stood up to any prison camp.

These brick apartments were built to house dock labourers. Walls wore layers of graffiti, where faded communist slogans outnumbered swastikas by at least five to one. Marc stopped in an alleyway filled with stinking pig bins[6] and glanced about furtively before pulling off his jacket and ditching it in an empty metal can.

He immediately saw a problem: his gun had been invisible below the jacket, but it bulged obviously when tucked in his trousers and wasn't much better when he put it in the cloth food bag. He'd gone less than thirty metres when he spotted the solution – a man's jacket hung tantalisingly out of reach.

Marc slowed down and checked the next couple of alleyways, eventually finding a stick. He then doubled back and gave the jacket a good whack, expecting it to drop down. Instead, he sent the entire line swinging across the street.

A shout of *thief* came from high up as he ripped pegs off the jacket and a spare shirt for good measure. Marc raced off expecting hot pursuit, but there was nothing behind when he looked over his shoulder. After a couple of turns he dropped his pace and put on the grey jacket.

A bunch of alley kids eyed him crossly as he stepped through their football game, but moments later Marc was back on the main drag, sweaty but unscathed.

---

[6] Pig bins – large metal bins into which people tipped waste food such as potato peel. The resulting slop was collected and fed to pigs.

The rest of the walk to Central Station was uneventful, but while the prisoner jacket had given Marc a clear identity he felt less confident in civilian clothes. His near-shaven head didn't suit a civilian and he decided to steal a cap first chance he got.

The route Marc had chosen involved starting at Frankfurt's colossal Central Station, which was the busiest in Germany. This was riskier than boarding a train at a smaller station, but Marc planned to buy a ticket to Leipzig, which was in the wrong direction for someone wanting to get back to France.

All being well, Marc would arrive in Leipzig late that evening, kill a few hours in the station and board the overnight Berlin–Paris express, arriving into Paris just before noon on Sunday.

But all wasn't well when Marc arrived at the station. First off, his train was due to leave in less than an hour and there were big notices up saying, *Due to increased security measures, please ensure you arrive a minimum of two hours before your train departs.*

Queues stretched from the station's triple-arched entrance, down a flight of thirty steps, then snaked several times across the courtyard. They were clearly hunting for Marc because the passengers were being sorted when they reached the station entrance.

Women, young children and older men passed through with nothing but basic document checks, but younger men and boys older than about ten were fed on to a separate section of the concourse, where Gestapo officers sat at tables, searching bags and asking detailed questions.

It seemed completely hopeless. Marc considered a few wacky schemes: sneaking into the station somehow, or dressing up as a woman. But even if he could pull something like that off, there was no way he'd do it in time to catch a train that left in an hour.

As well as the Gestapo officers on desks, Marc was sure there would be plain-clothes officers nearby, whose job was to keep an eye out for people like him who backed off when they saw the heavy security. He used his espionage training, being careful not to stop or to make it too obvious when he looked around.

'Support the front?' someone asked.

Marc found himself face to face with an old granny draped in swastikas. She had a wooden tray strung around her neck, which was filled with cheap-looking tin swastika badges.

The old girl broke unenthusiastically into a prepared script. 'All good Germans must contribute. Our men fight heroically, but everyone must pull together for ultimate victory. This badge is a symbol of solidarity of German people . . .'

Marc had to make a rapid decision. Although his German was good, he spoke it with a distinct French accent. The more words he spoke the more his accent became clear, so he raised a finger and said, 'One,' as he rummaged in his pocket for coins.

This was the problem with being out on the street. You could sit up on a roof, poring over maps and timetables, planning every detail, but out in the real world things got unpredictable. You faced rapid decisions, when one wrong answer could earn you a bullet through the head.

Marc felt jittery as he moved away from the station, pinning the tin swastika to his freshly stolen jacket. He looked back for any sign of a tail and took a couple of side streets, before doubling back to be completely sure.

He had to decide between returning to Großmarkthalle or trying to see if the security situation was any better at one of Frankfurt's smaller stations.

Without a map and afraid to stop and ask for directions with his French accent, Marc walked for almost thirty minutes. He'd started at Großmarkthalle in the east, passed through the city centre and was now travelling west.

Shops and cafes closed at lunchtime on a Saturday and it was a lonely walk through a prosperous part of town.

Although Marc felt stronger than he'd been when he first escaped, months of prison rations followed by his recent stomach bug meant he was a shadow of the boy that had passed espionage training a year earlier. His big worry was that if he kept walking for much longer, he'd crash before making it back to Großmarkthalle.

It felt like fate when a bus approached, with *Höchst Station* on the destination board and a bus stop less than a hundred metres ahead. Marc only knew three things about Höchst: it was roughly 10 km further west on the city's outskirts, he'd seen it mentioned as the stopping point for many trains going west when he'd studied timetables, and lastly it had several large chemical plants which were dreaded by prisoners.

Marc wasn't familiar with German buses, and felt like everyone was staring as he found a seat near the back next to a pregnant woman, then fumbled to pay the conductor.

The bus swept through quiet streets, then beyond the edge of the city proper into the industrial belt. The route was timed to coincide with a shift change at a big factory and the bus filled to bursting with matching brown overalls and the smell of men needing a wash.

Marc was intrigued by two workers standing a few rows ahead. They asked the conductress why the bus was late and whether they were likely to make up enough time to connect with the last train to Mainz.

Marc's German geography was far from perfect, but he'd picked up some knowledge while working for Commandant Vogel. Höchst was ten kilometres west of Frankfurt; Mainz was another fifty kilometres in the direction of the French border.

It took twenty minutes for the bus to reach Höchst. As Marc only knew it for chemical plants, he was surprised to find his bus driving through narrow streets, with picturesque medieval buildings like something out of a Nazi propaganda film.

Höchst station felt oversized for a modest town and the atmosphere couldn't have been more different to Frankfurt Central. Half the platforms were overgrown and there were more people queuing to use the telephones than waiting to board trains. Most importantly, there were no Gestapo blocking the main entrance.

The two men who were anxious about connecting with their train rushed to be first out of the door when the bus pulled up.

'Sorry,' the lead man said, doffing his cap as he squeezed past another passenger. 'If we don't get that one, it's another two hours.'

Marc liked the idea of putting another fifty kilometres between himself and Frankfurt and decided to make a run for the train himself. He raced into the station close behind the men, then up on to a footbridge to get to the platform.

The two men passed an inspector waving tickets, but Marc didn't have one and came to a halt. He couldn't double back because the ticket inspector had already seen him and it would have looked suspicious.

'Can I buy a ticket on the train?' Marc asked, doing his best German accent. 'I've got to get home to Mainz or my mum will *murder* me.'

There was a second railway worker on the ticket barrier. He had eyes that bored straight through you, and Marc was in enough of a state even without freaky eyes staring at him.

'Mainz, one way?' the guard asked, as he opened a leather waist pouch filled with ticket rolls.

'Yes,' Marc said, eying the train anxiously as he pulled out a ten Reichsmark note.

'Payday, eh?' the guard laughed. As he tore off a pink ticket, the guard pointed to the whistle hanging around the freak-eye's neck. 'Don't worry, son. No one's going anywhere until my friend here plays a tune. You enjoy your weekend.'

'Thank you, sir,' Marc said.

A couple more passengers belted past the inspector waving tickets as Marc pocketed his change and dashed across the platform. The third-class carriage was jammed with bodies and Marc shuffled inside, reaching anxiously for a grab-handle as the train pulled out.

CHAPTER FOURTEEN

The Mainz train was a stopping service that took two hours to go fifty kilometres. Marc studied the timetables in the ticket hall when he arrived and got a shock when he found a station attendant staring over his shoulder offering to help.

While Marc was wary, she identified his accent and seemed pleased by an opportunity to try out her French.

Marc explained that he was heading home, but had boarded the wrong train. The woman said he could pick up a Paris-bound train at the border town of Saarbrücken, but the last train to Saarbrücken had left for the day and the next wouldn't run until Monday. If he didn't want to wait, he could either get a train back to Frankfurt, or a bus to Saarbrücken.

Marc found the bus station. Buying the ticket to

Saarbrücken was no problem, but the bus didn't leave until six the next morning. There didn't seem to be much security around, but Marc doubted he'd get through a whole night without a bored policeman or railway official sticking his nose into his business.

It was starting to get dark as he wandered through the area around the station. He had food, but needed a place to spend the night. The weather was mild, so a field or shed would do, but the area around the station was built up and without a map he was worried about straying too far and getting lost.

Marc finally caught a break when he turned into a street of modest houses. An elderly couple were stepping out of their front door, arms linked and dressed to impress, with dancing shoes on their feet. What hooked him was the way the man walked around the side of his house to check that all the windows had been closed, making Marc sure there was nobody else home.

Breaking in was a risk, but Marc was still weak from the stomach bug. Two hours standing on a hot train followed by a long walk had left him with painful feet, banging head and a desperate thirst. A few hours in an unoccupied house would enable him to sit down, use the bathroom, take a long drink and possibly even heat some food.

He gave the old couple time to clear off before doubling back. All the houses in the street were identical:

front gardens had well-trimmed hedges and the main doors were all at the side of the houses, giving a degree of privacy to anyone breaking in.

The front door had a deadlock, making it impossible to pick without locksmith's tools. Frustrated, Marc crept around to the back garden, keeping low so that his head wasn't seen over the dividing wall.

The door opening on to a large back garden had no keyhole, just a knob that turned from inside. Marc shoved one of his cook's knives into the gap between door and frame, then used it to push back the spring-loaded latch bolt holding the door shut.

Marc had lived in institutions his whole life and always felt envious when he entered a proper home. It was a cosy little place, with a rocking chair, framed pictures and a collection of pipes hung from the wall.

As Marc grabbed the door handle to explore the rest of the house he heard a scratching sound. He glanced back, making sure his exit was clear, before clutching the cook's knife and opening the door into a hallway.

A floppy-eared spaniel scrambled to the door as it came open. When it got a sniff of stranger, the little dog yelped, skidded on the polished floor, then backed up to the bottom of the stairs and started growling.

'I'm not gonna hurt you, boy,' Marc said softly as he stepped into the hall.

He needed to be sure nobody else was home, so he

quickly confirmed that the kitchen and dining room were empty, before heading upstairs and checking three more rooms – the couple's bedroom, a tiny study and a neat space that looked like it belonged to a boy who'd grown up and left home.

The dog thought exploring the house was a great game. He chased Marc, and when they reached the bottom of the stairs again the spaniel started jumping on to Marc's leg and slobbering over his wrist.

The rough tongue tickled and Marc couldn't help laughing, but the dog's attention made Marc emotional. He realised it had been months since he'd felt any kind of affection, human or otherwise.

Marc didn't want the break-in reported to the police before he left town, so he made a mental note of where everything was in the kitchen before striking a match and putting a kettle on the gas stove. It was nearly dark, but the kitchen overlooked the street, so he left the light out as the water boiled.

Finding nothing but revolting ersatz coffee, Marc settled for hot water. He sat in the rocking chair in the back room, pulling off his sweaty rubber boots and letting steam from the mug rise up to his face. He munched some of his biscuits, then unwrapped a big piece of sausage and shared it with the dog at his feet.

Marc reckoned the old couple would be dancing for at least a couple of hours. He was tempted to just relax in the chair, but he'd been intrigued by his glimpse of the grown-up-and-left-home bedroom upstairs.

The dog stayed back, licking biscuit crumbs off the floor as Marc headed upstairs. The bedroom was tiny, with one wall steeply angled by the roof. There was a mirrored dressing table and a poster of a Mercedes racing car on the wall above a single bed.

Items were set out in a certain way, giving the impression of a museum exhibit. Marc realised that the room's owner wouldn't be coming home when he noticed a small wreath of dried flowers resting on the neatly-folded soldier's uniform at the foot of the bed.

Everything had been preserved, including a pocket watch in the bedside drawer and polished boots under the bed. As well as the dead man's clothes, the wardrobe contained items from childhood, including a Hitler Youth uniform, shirts and trousers, plus pairs of good-quality shoes, going all the way back to a size that would fit a lad of eleven or twelve.

After all Marc had been through, he'd have happily lobbed a live grenade into a truckload of Germans, but he could almost feel the dead soldier's presence, and the idea of stealing his things unsettled him.

However, Marc's rubber boots, tattered clothes and shaved head made him look very different from a typical German fourteen-year-old. Regular shoes and German clothes would help him fit in and he guiltily tried on a couple of pairs of shoes, seeking the best fit, before grabbing a cap, trousers a spare shirt and a couple of sets of clean underwear.

Marc would have to leave the house before the old couple got back from dancing, but a glance out of the window made him wonder about the possibilities of sleeping in the back garden, rather than setting off on another risky trek through town.

Like most lawns in Germany, this one had been dug up to grow food. Tall rows of beans and crude glass frames shielding rows of tomatoes made plenty of hiding spots. As long as Marc hid his break-in well, he saw no reason why the middle-aged couple would come back late from dancing and start rummaging around in their garden. And even if they did, Marc felt he'd have a better chance in a confrontation with them than if he got picked up by cops wandering the streets, or sleeping on a bench near the bus station.

*

Marc had fun playing fetch with the dog, but he let it carry on too long and had to rush out of the back door when the couple got home at 10:30. If spaniels could talk Marc would have been in trouble, but the couple just seemed slightly baffled by the excitable state of their dog.

'He's panting,' the woman said. 'You mad thing, it's after your bedtime!'

Marc had already put his things outside and made a

little den by spreading some cloth sacks he'd found over the damp earth between fruit bushes and the garden's back wall. The ground was lumpy and he worried about rain, or sleeping late and missing the bus.

He'd only managed a few short naps, a stare-out competition with a ginger cat and hours of dark thoughts when the sun began rising. The garden was too dangerous in daylight and the dead soldier's pocket watch told him it was just before six as he crept around to the alleyway at the side of the house.

He brushed mud off the clothes he'd slept in and stashed them in his barley sack as he changed into shoes, shirt and trousers stolen from the wardrobe inside. Marc heard the lady of the house stirring as he laced his shoes. His heart ran quick as he grabbed his things, jogged up a path and tried his best to silence a squealing front gate.

Marc didn't want to arrive at the bus station early, because the more time you spend hanging around the more chance there is of someone spotting you and asking questions. But he'd badly underestimated the time he'd spent walking away from the station the night before and ended up running through dead Sunday morning streets.

The dark blue bus was half full. All but one of Marc's fellow passengers came from a group of newly-trained soldiers returning from leave, none more than eighteen years old.

He half listened to bold claims about things they'd

done to their girlfriends, along with more fearful conversations about where they'd be posted.

With room to spread out, Marc dozed off after his restless night. He was woken by a shout, and absolutely crapped himself as his blurry eyes saw the bus being stormed by an army patrol.

Two military police officers were working from the front of the bus, inspecting documents, while a third stood in the road behind. Gun poised, in case anyone tried doing a runner through the emergency exit.

Marc's documentation was authentic. But it was for a French labourer returning home from Frankfurt on compassionate grounds, which made his presence on a bus between Mainz and Saarbrücken grounds for suspicion.

He couldn't understand why the young soldiers looked so worried as their identity documents and leave passes were inspected. When the military police officers got to the row in front of Marc an out-of-date document was discovered.

'This counts as desertion,' the military policeman shouted, as the youth got dragged off the bus towards a waiting army truck. 'Deserters can be shot!'

Marc watched nervously as the deserter was slammed against the side of the truck, frisked for weapons and then beaten down ruthlessly with a baton swipe across the back of the knees.

'So,' the military police officer said, when he reached Marc. 'You're a little young to be my concern just yet, eh?'

Marc had his paperwork ready, but the officer didn't ask for it.

'Did you see any soldiers leaving this bus after they boarded?' he asked.

'I've been asleep some of the way, sir,' Marc stuttered. 'But I don't think so.'

The military policeman pointed at the young troops. 'And what has this rabble been talking about, exactly? What are they up to?'

'All I heard was rude things about their girlfriends,' Marc said.

The military policeman roared with laughter and some of the young troops even joined in. But Marc's laughter got pricked by the next question.

'Interesting accent,' the policeman said. 'French?'

Marc's tongue suddenly felt huge and he didn't fancy admitting to being a foreigner in this company.

'Germanisation,' he said, thinking back to what Commandant Vogel had told him. 'I was born in France, sir. But my parents died and now I'm here.'

The officer turned to his colleague and laughed. 'Blond hair, blue eyes. Proper little Aryan trooper, isn't he?'

'Needs feeding up,' the colleague replied.

Lying about Germanisation could have backfired if Marc had been asked for his documents, but he'd correctly judged that the military policeman was fixated on his hunt for deserters. The officer nodded and moved his attention to the soldiers in the rows behind.

When the inspection was over and the bus moved off, Marc was surprised by a hand tapping his shoulder. One of the young soldiers held out the stub of a paper tube, with three boiled sweets left in the bottom.

'Take 'em,' he said. 'Thanks for keeping your mouth shut.'

Marc realised he'd slept through something. Exactly what it was eluded him, but it was interesting that these young troops were prepared to jump off their bus and risk being shot as deserters sooner than go east to fight Russians.

CHAPTER FIFTEEN

After a four-and-a-half-hour ride, Marc worked the stiffness out of his legs on a long walk, from the bus station at Saarbrücken's edge to the train station in the city centre.

The town was a transport hub for cargo coming from France, and for the vast coal fields in surrounding Saarland. British bombs hadn't stopped much cargo moving, but there was more sense of a war going on than in Frankfurt. Many windows were sandbagged or shattered and streets were scarred with bombed buildings and fire damage.

The station also served as a border post, so it was no surprise to find tight security. The guards at the station entrance stood rigid as Marc passed unhindered on to a small passenger concourse. He sat on a wooden bench, studying the set-up.

Trains travelling within Germany were boarded in the normal fashion, but there were special signs guiding passengers wanting to enter France to a customs post. Marc had hoped he'd be able to buy a ticket to somewhere in Germany, pass through the barriers and then sneak aboard a train travelling into France. But apparently he'd have to clear customs.

After using the toilet, Marc strode casually towards the entrance to the customs post. From a distance he'd been pleased by the lack of a queue, but as he closed in he saw that the gates were locked.

'Looking for something, son?' an elderly station cleaner asked.

'I like watching trains,' Marc said, slurring his voice in the hope that it made him sound simple instead of French. 'I wanted to see a French train.'

The man opened his mouth, revealing nothing but gums as he pointed to the back of the station. 'Best place to watch trains is up on the road bridge over the tracks. It's quite a sight when the Frankfurt–Paris express goes through at full steam, but you won't see it today. The border doesn't reopen until midnight and the first passenger train runs a little while after that.'

Marc kept up the thicko act and looked suitably crestfallen. 'Oh . . .'

It was just past two, so Marc had ten hours to kill. He reckoned his best bet was a cinema he'd passed a few hundred metres before the station.

The screen showed a rolling programme of news, short films and a main feature which was advertised as the *sensational retelling of a romantic German folk tale*. After a short queue, Marc entered a room with a cavernous screen that was about four-fifths full.

An usher fought through a haze of cigarette smoke and showed him to a seat in a quiet corner off to one side. The main feature had about twenty minutes to run, then there was a short interval before the news started.

The lead story was an optimistic take on progress at the Russian front, with photos of advancing German troops, suspiciously clean tanks and shots of raggedy Red Army soldiers being dragged away in leg irons, where *these barbarians would be re-educated in the ways of the Reich through a corrective programme of physical labour*.

The next story told how a group of Berlin Jews had been executed for forging clothing coupons, while the third and final story was a lighter piece, following the life of teenage German girls working in a munitions factory.

The factory was even cleaner than the tanks fighting in Russia and of the many factory workers Marc had seen over the past months, these were the only ones whose uniforms showed cleavage.

The report ended with shots of these attractive German girls playing volleyball on a beach in frilly swimming costumes, and Marc wasn't the only teenager in the auditorium who had to fend off an erection when the lights came on and everyone had to stand for the German national anthem.

Although there was always a big changeover of patrons during the intervals, the programme was continuous. There was no restriction on how long you could stay, though to avoid the place turning into a doss house there were signs warning that the management would kick out anyone who fell asleep.

Marc fought heavy eyes as he watched the cycle of news, short films, adverts and main feature twice, then perked up when the news came on for a third time and he got to see the volleyball girls again.

He took a good drink from the water fountain in the cinema foyer, then stepped out bracing himself for a final showdown at the customs post in the station. Seven hours in a cinema had given him more than enough time to fret about his chances.

After a brief panic because everything looked different in the dark and he couldn't remember which way the station was, Marc set off at a slow stroll, unaware that eight lads in Hitler Youth uniform were tailing him.

'Who are you?' a stocky lad of about thirteen demanded, as he jogged up alongside. 'It's a Sunday. Why aren't you in uniform?'

'I'm not from around here,' Marc said, copying the

questioner's surly tone. 'Not that it's any of your business.'

A couple of bigger lads ran ahead, blocking Marc's path.

'So you're in Hitler Youth?' one of them asked.

'I am where I live,' Marc said, not comfortable with the lie, but wanting to admit to being a foreigner even less.

'And where's that?'

'Mainz,' Marc replied.

Apparently this wasn't a good answer. The boys moved in closer and jeered. One spat on the cobbles. Marc thought his accent had given him away, but the outrage had nothing to do with his voice.

'Mainz,' the big lad spat. 'Didn't we beat you lot down at our last training camp?

'How *dare* you come down here and pollute our turf, Mainz boy,' another lad added.

The bulky pair pinned Marc against a shop front. Up to now he'd assumed the Hitler Youth were a brotherhood who persecuted foreigners and Jews, but apparently they persecuted each other too.

'Strutting around *our* town. How dare you!'

'What's in his bag?' Another lad asked.

'We should strip him down and piss on him,' a kid at the back who was no older than ten suggested.

Another boy spat in Marc's face. 'Dirty little gypsy.'

'My dad's waiting for me at the station,' Marc said weakly. 'He'll come looking in a minute.'

Someone tugged at the sack slung over Marc's back, but it had his change of clothes, his travel permits and the gun inside, so it would take more than that to get hold of it.

'Cheeky,' the boy said. 'Give us the sack before I slap you one.'

'Give him a kicking, then throw him in the canal,' the little kid with the bright ideas said.

'What if he can't swim?' someone asked.

'Is that our problem?'

Marc wished he had one of the cook's knives in his trouser pocket, but all that came to hand was the uncoiled spring he'd been using to pick locks. He knew he'd only get one surprise move before eight bodies overwhelmed him, so he waited until the kid moved close to spit in his face again before launching an upwards punch.

The spring speared the base of the kid's jaw, and on through his tongue. Blood spurted as Marc ripped the spring out. In the darkness most of the other kids thought it was a knife and backed off.

One of the kids pinning Marc to the wall stood his ground. Marc braced against the shopfront for extra strength as he kneed him in the groin. There was now a gap and Marc made a run for it. A fist glanced his head but he broke free.

Marc still had the bloody spring in his hand as he started running. Some of the gang stayed back to look after the kid who was bleeding, but three lads – including a big bugger who looked about fifteen – gave chase.

Marc ran four hundred metres along a busy street. There were loads of adults out for a Sunday evening stroll, but none of them thought they were seeing anything more serious than teenagers messing about.

By the time Marc cut into a side street, the biggest lad had mercifully dropped back and the only kid close enough to matter was the un-kneed half of the team who'd pinned him to the wall. He was fast and managed to get an arm around Marc's neck and bundle him hard into a wall.

'Mainz shit,' the lad gasped, as he pulled a dagger from his belt and lunged at Marc's belly. 'I'll kill you.'

Marc's mind flashed back to his training on CHERUB campus. Catching the clumsy lunge was ludicrously easy. Marc twisted the kid's arm behind his back, making the blade drop to the cobbles. The kid still had forward momentum from the lunge and Marc used it to shove him head first into a metal gate post.

The hollow thud was awful and Marc couldn't help thinking that he'd cracked the kid's skull. He was groaning horribly as Marc jogged out into a wider street, where there was a bit of moonlight.

Marc had a splash of the stabbed kid's blood across the entire width of his shirt. Running attracted attention, so after half a minute and a turn into another alleyway Marc dropped his pace to a brisk walk.

The alleyway exited within sight of the station, though he'd found it more by luck than skill. Remembering the rear exit that the toothless station guard had pointed out earlier on, Marc walked a long footbridge that spanned more than twenty sets of tracks.

When he reached the rear of the station, Marc found a tiny row of shops that looked like they'd been shuttered for years. After glancing about to make sure nobody was around, he backed into a urine-scented gap between a wall and a disused flower kiosk

Buttons flew off as he ripped away the bloody shirt. After using it to wipe blood off his hands, he stepped out of the shorts, then opened his sack, then grabbed a clean shirt and a brown cap which he pulled down over his face.

A couple walked by arm in arm, and Marc kept still in the shadows. When they'd passed without seeing him he sorted his backpack. There was every chance he'd be searched by customs, so he ditched the gun, the bloody spring, bloody shirt and the larger of the two cook's knives.

The pocket watch said it was only a quarter past eleven, but his paperwork was immaculate and he fancied his chances queuing on a station concourse better than his chances if the local Hitler Youth caught up with him on the street.

When he reached the station, Marc was surprised to find the customs desk already open.

The attendant waved him through. 'Have your ticket and passport, or your travel warrant, ready when you reach the top of the stairs.'

Marc felt queasy as he climbed two dozen wooden steps, then queued behind four other passengers in front of a wooden counter.

The walls were decked with posters, warning about everything from not discussing secret information on busy trains, to a list of items that civilians weren't allowed to carry into occupied territory. In the best Nazi tradition, smoking in the customs line earned up to ten years in prison, while everything else was punishable by death.

Marc waited long enough to get seriously nervous. He was sure his description would have been sent to all French border crossings, but he had no idea how significant this was.

Were the customs officers looking out for a few faces? A few dozen? Or thousands? Had they tracked down and distributed his photo, or were they working from a vague description like the one he'd seen in the newspaper? Did they have information about the distinctive X-shaped scar above his eye? And were they aware that he'd been able to gain access to original travel documents?

When it was Marc's turn, the customs officer kept staring at the lookalike photo, then up at Marc's face. He fought off a shudder every time she frowned.

'When was this photograph taken?' she asked, in French.

'About a year back. I've lost weight.'

'You look younger too,' the woman said suspiciously, as she tapped the warrant with the end of her pencil.

Marc smiled uneasily. 'Better than getting older, I guess.'

'This warrant is dated yesterday. Why have you taken so long to get here?'

'I got on the wrong train,' Marc said. 'I ended up in Mainz.'

'And why are you boarding here? Wouldn't it have been quicker to go back through Frankfurt?'

Marc felt doomed. The woman seemed to have it in for him.

'I thought it would be better to keep moving west.'

She rolled out her bottom lip. 'There are no trains from Mainz to Saarbrücken on weekends.'

'The lady at Mainz told me to get the early morning bus,' Marc explained.

'I guess that's in order,' the woman said, but still didn't seem sure. 'What were the grounds for your compassionate release?'

'My mother died. I've got three little sisters who need looking after.'

'That sounds like a handful,' the woman said, her tone warming as she smiled slightly and placed a rubber stamp on the travel warrant. She then pulled out a small cardboard exit visa and wrote Marc's false name on it before stapling it to the warrant.

'This warrant entitles you to third-class travel only. Seats are on a first come first served basis, but you're early so at least you'll be near the front of the queue. The train should arrive at 00:25. If it's too busy to board, the next one is at 04:30. Be sure to keep your documents ready for further inspection on the train, and when you arrive in Paris.'

The woman slid Marc's documents across the desk.

'Is that it?' Marc asked.

'That's it,' the woman said. 'Have a safe trip and I hope it all works out with your family.'

CHAPTER SIXTEEN

Marc entered France squashed on a wooden bench, with a fat German bun-head jabbing him with her elbows as she click-clacked a set of knitting needles. But he was still luckier than the people forced to stand for a six-and-a-half-hour journey, which ended up taking nearer ten.

There was a ninety-minute wait to link up with two carriages full of German soldiers at Metz, a halt for a 'surprise' additional customs check that regular passengers said happened every trip, and finally delays outside Paris because they'd arrived in the morning rush hour and there were no free platforms at the station.

Marc felt a kind of cautious elation as he looked out the window, noticing all the little details that differentiate one country from another: French vehicle licence plates, French road signs, types of houses, shops and clothes.

It looked like home, but Paris Nord Station was a stark reminder that France was under German occupation. There were more German uniforms here than anywhere Marc had been in Frankfurt and the French passengers were made to stand at a security barrier for twenty minutes while soldiers and then German civilians got priority.

Only a tiny proportion of passengers were searched by the Gendarmes[7] at the head of the platform and Marc had no problem.

He'd grown used to Frankfurt, where civilians moved about freely, but he only got as far as the station exit before encountering another checkpoint. This time everyone was expected to flash their identity cards, but Marc's prisoner transfer documents didn't pass muster and he was taken aside and made to sit in a small waiting room.

He spent a nervous fifteen minutes, sitting between a towering Belgian with a sack of tools between his legs and an exhausted-looking woman who'd arrived from Germany without documents for her newborn baby.

Marc was called into an office by a bored-looking Gendarme, who took Marc's latest false name and put

---

[7] Gendarme – a French civilian police officer.

stamps on several pieces of paper, before directing him to travel four stops on the Metro to an office where he could register for his new identity and ration cards.

The last time Marc rode the Metro was two summers earlier, just before the German army arrived. The ticket offices now accepted Reichsmarks, but gave change in less valuable French francs. To save electricity the Germans had removed all the light bulbs from Metro carriages, so Marc had the eerie experience of rattling through underground tunnels in the pitch-dark.

Marc expected to spend hours queuing for his new identity, but the process was remarkably efficient. After a short queue, his photo was taken, then he was asked a few simple questions. His thumb was inked and a thumbprint added and by the time the blank identity card, employment status card, tobacco entitlement card and seven-day emergency ration card had been filled out a man had emerged from the darkroom with his freshly developed photograph.

'You'll need to register locally for a permanent ration card and get your identity card updated once you have a fixed address,' the woman explained. 'As you're over fourteen, you're required to register for work at a Labour Administration office by Friday at the latest.'

She adopted a more friendly tone before continuing in a quieter voice. 'If I were you, I'd try sorting out a job before you register for work. If you turn up and say you're unemployed, they'll send you straight to the truck factory on Ile Sequin.'

'I'll keep that in mind,' Marc said gratefully.

Marc's new identity was Michel DePaul, fourteen years old. He'd told the clerk he was born in Dunkirk. The record office there had burned to the ground during the evacuation of British troops two years earlier, making it impossible to trace a birth certificate.

When he hit the pavement outside the office, Marc felt like making some big, triumphant, fist shaking, *I've done it*, gesture. But this would have attracted attention, and the moment coincided with a bicycle fitted with a number plate whizzing past.

The thought of number plates on bicycles amused Marc, but it was another sign of the tight lid the Germans now kept on their French subjects. And it would make life way more difficult if he ever needed to steal a bike.

*

Marc's position was stronger than when he'd been creeping around Frankfurt two days earlier, but his ultimate goal to rejoin his espionage unit in Britain was still a long way off.

Marc planned to settle in Paris for at least a week. There were dormitories where working men could pay by the night, and with over half of the working-age men in France held prisoner in Germany, he thought it shouldn't be too hard to find a job.

Security was tight. But Paris felt no more overbearing than what Marc had experienced in the rest of German-occupied France, and he reckoned it would be easier to go unnoticed in a city of four million than in a small town or village.

Marc had given a lot of thought to what he'd do after he arrived in Paris, both before his original failed escape, and over the past few days. It boiled down to three options:

(1) Travel back to Lorient and try making contact with the espionage circuit he'd been working for when he'd been arrested eleven months earlier. Marc had once favoured this plan but deeper thought unearthed serious flaws:

Firstly, getting to Lorient required a special permit to enter a restricted military zone. Second, Lorient was the first place the Gestapo would go looking for him. Third, eleven months is a long time in espionage terms and Marc might arrive to find that his circuit had disbanded, or been rumbled by German agents.

(2) Travel south into Vichy France and try making contact with friendly officials at the American embassy. If this failed, keep moving south and try crossing the border into Spain.

A special pass called an Ausweiss was required to cross into Vichy France, but thousands of people crossed from north to south every day and getting a pass, or sneaking across the border, wasn't impossible.

However, America and Germany had declared war on each other while Marc was in prison and he had no idea if the American embassy in Vichy had been closed down.

If he had to escape into neutral Spain, it would involve a dangerous foot crossing over the Pyrénées mountain range and Marc had no idea how tough the border security was.

(3) Finally, Marc could try making contact with a Paris-based resistance group. If they had a radio transmitter they'd be able to get a message to British intelligence, saying that he was safe and asking for instructions on the best way to get to Britain.

Marc decided to pick this third plan, but it relied on patience and a degree of luck to unearth resisters who were trying not to be found. Plus, any resistance circuit with decent security procedures would be extremely wary when Marc approached them and told an unlikely sounding story about being a fourteen-year-old British intelligence agent.

CHAPTER SEVENTEEN

Marc queued twenty minutes for fresh bread that cost three times what he'd expected it to, then found a street he remembered with a few dormitory houses and a half reasonable one with beds to rent.

On Marc's first visit to Paris he'd been unable to stomach the stench and filth of a working man's dorm. But even the grottiest dorms had flush toilets, single beds with sheets and mattresses, electric light, and other facilities that felt like paradise compared to a prison hulk in Germany.

Unfortunately, inflation had also taken its toll on the room rate. Marc didn't have enough money to cover one night and the landlady looked like she was going to bounce him head first down the front steps when he asked for credit.

Needing some fast money, Marc found a pawn broker's shop and tried to sell the pocket watch. It was a nice piece, but a message engraved in German meant the broker turned up his nose.

'How's a French boy end up with a watch like that?' he asked, running a hand through slicked-back grey hair.

Marc realised he should have prepared a story. 'Guy my aunt was sleeping with,' he stuttered.

'You stole it off the bedside table?'

'It's not stolen. He gave it to her.'

The pawn broker laughed and slid the watch back across the counter. 'Funny thing to give to a lady. You know what the Germans will do to me if they catch me with a soldier's stolen watch?'

'I think it's silver though,' Marc said. 'I need to get back on my feet. Enough money to pay for a dorm room for a couple of nights, that's *all* I'm asking for it.'

But offering to sell cheap only confirmed the pawn broker's opinion that the watch was stolen.

'You're lucky I don't call the Gendarmes on you,' the broker said, as he pointed at the door. 'I think it's time you left.'

Marc backed out, and walked the next couple of streets at a brisk pace, just in case the broker decided to cause trouble. He didn't think he could face another pawn broker, so over the next couple of hours Marc pounded pavements, stopping in every bar, cafe and restaurant asking if they wanted a waiter or a cigarette boy.

One place offered a meal if he scrubbed a vast mound of pots, but there was nobody paying cash. The long walk and warm weather brought Marc out in a sweat and for the first time since he'd been shaved in hospital, he felt the familiar itch of body lice partying under his arms.

Marc had only slept in short bursts on the overnight train, so when it got to 6:30 he was struggling to keep his eyes open and his ankles were seizing up. Paris had a 10 p.m. curfew for any French person not travelling to or from work.

It seemed ludicrous for Marc's escape to go awry because he was short a few measly Francs, but his prison camp in Frankfurt had been full of men scooped up on minor charges such as drunkenness, urinating in public, or breaking curfew.

With no realistic prospect of landing a job and getting paid within the next three hours, Marc no longer had the luxury of trying to get money by honest means. His friend and fellow agent PT Bivott had taught him how to run card tricks, but Marc didn't have cards and you needed regular practice. This left thieving as his only option.

Marc was starving, so he entered a crowded cafe and exchanged two precious ration tickets and most of his remaining cash for a meal of carrot and celery soup, followed by a titchy piece of white fish served on a bed of green beans and potatoes.

It was the first proper sit-down meal he'd eaten in ages, and two glasses of red wine gave him courage as he watched a pair of busy waitresses going back and forth, stuffing money in the till. His plan was simple: leap over the counter, open the cash drawer, grab a bundle of money and run like hell.

Marc was all set, and even counted *three, two, one, go* in his head. But he was still rooted to his chair fifteen seconds later when the waitress asked if he'd enjoyed his meal, and whether he'd mind moving to the bar because there was a queue of patrons waiting for tables.

'It was very good,' Marc said, as he drained his glass of red.

The bold plan suddenly felt less wise: it was a busy area, and with so many customers around it only took one person to stick a leg out and trip him, or one determined man who could outrun a fourteen-year-old with tired legs.

Marc realised the booze had almost made him brave enough to try something suicidal as he sauntered off along a street of cafes and bars.

He was disgusted by the German soldiers and airmen with dolled-up French girls on their arms and wished these women could see the conditions their brothers, fathers and husbands were enduring in camps a few hundred kilometres west.

A frail creature caught Marc's eye. She was stepping out of a cafe, with a fat waiter making a big fuss over her, draping a fur cape over her back and begging the old girl to come back soon. She had pearls on her neck, and rings with diamonds the size of ladybirds.

Marc felt sure there'd be plenty of Francs in her black leather bag and followed her away from the main drag down a side street. There was hardly anyone about and all it would take was a good shove to knock her down and a tug to rip the bag off her shoulder. But Marc had been knocked about all his life, by teachers, by thugs like Fischer, by his orphanage director and a Gestapo officer who'd ripped his front teeth out.

He'd seen sadism in little kids at the orphanage – the ones who stamped on baby ducklings and bent weaker boys' fingers back for the hell of it. Although Marc had killed ruthlessly when under pressure, he had a strong sense of right and wrong and ripping off a little old lady turned him into all the things he was fighting against.

So Marc let her toddle down into a Metro station and walked back towards the action. For the next hour he wandered. It was almost nine. Bars and restaurants were winding down as French people headed off to beat the curfew.

Marc decided to use his last coins to buy a beer in a near-empty cafe. The staff were getting ready to close up. Instead of the risky snatch-and-grab he'd planned earlier, he decided on a sneak raid behind the counter as the staff mopped the floor and stacked chairs on tables.

He was about to enter the cafe when he noticed a German officer, sitting outside a restaurant with his coat draped over the back of a chair. He was leaning forward, with his attention fixed on the teenager he was groping, while his leather wallet poked from an inside pocket practically begging to be pinched.

It felt like fate. There were plenty of other people in the street, and Marc had to make sure nobody was looking, without it being obvious that he was making sure nobody was looking.

After a deep breath, he approached the back of the chair. At the last second the officer broke off from the teenager, but Marc was committed. The girl saw exactly what happened and looked around, clearly shocked.

'You OK?' the German asked.

Marc expected the teenager to start shouting, followed by the German standing and pulling a pistol on him. But the girl – who looked no older than sixteen – just gave an awkward smile and slapped the side of her neck.

'I got bitten,' she stuttered. 'Insect, or something.'

Marc couldn't believe his luck, but he was so scared that everything was going in slow motion and his limbs felt like wooden stumps.

The German laughed noisily. 'If you've got fleas I hope you don't give them to me.'

Marc pushed the wallet into his trouser pocket, walking as fast as he dared without it looking dodgy.

When he got into a side street and checked nobody was behind him, Marc backed up to the wall. He was shaking with pure terror and hated the fact that despite all the espionage training he'd been through, his survival was down to pure dumb luck.

CHAPTER EIGHTEEN

The officer's wallet was a good score. It left Marc with enough French Francs to get by for two or three days, and German Reichsmarks for a few more after that. He abandoned the wallet in a dark Metro carriage, while riding five stops back towards the dormitory he'd scouted earlier on.

A fifteen-minute queue to get papers checked coming out of the Metro station brought Marc precariously close to ten o'clock curfew, though it was a nice feeling when the German handed his identity card back with a polite, 'Thank you, move along.'

Marc banged on the door of the dormitory house, but the woman he'd spoken to earlier would only shout through the letterbox.

'We're full. Can't you read the bloody sign?'

Marc gasped. 'But you said you had *loads* of room!'

'Half a symphony orchestra came and filled me up. Whoever pays first gets the bed.'

'But it's five minutes to ten. Can you let me in? What if I sleep on a chair, or in the hallway?'

'All my residents are in and the door's bolted,' the woman said.

The slamming letterbox flap was her final word.

Marc raced down the street, but apparently the symphony orchestra had filled up the other two dormitory houses as well, and now it was four minutes past ten. Based on what he'd heard in Lorient, people picked up after curfew generally got a couple of nights in the cells and a stiff fine, but he'd met enough curfew breakers in Germany to know that deportation wasn't out of the question.

'Have you got any ideas where I can go?' Marc asked, at the third place.

The man gave him directions to another dormitory house several streets away. There was no lighting because of the blackout, and the dead streets reinforced Marc's opinion that curfew was not something people took lightly.

He walked briskly, ignoring growing tiredness and aches in his legs. His heart thudded and he reached full-on panic when he realised he'd taken a wrong turn.

The fourth dormitory house was above a laundry. He

rang the doorbell and waited a full minute before trying a shout through the letterbox.

'Hello? Can someone help me, please?'

A first-floor window came open and a hairy-chested man with a cigar butt hanging off his lips looked down.

'If you don't stop that racket, I'll sling the piss pot over your head,' he shouted. 'Now sod off.'

Marc heard an entire dorm erupt with laughter at his expense.

'I need somewhere to stay,' he begged. 'The floor, the hallway. I don't want to get picked up.'

'That's not my lookout,' the man shouted. 'I've warned you once. If I have to come down there you'll be sorry.'

The window slammed and Marc turned around, thinking about luck: one minute a pretty girl turns a blind eye and saves your life. Half an hour later a symphony orchestra turns up and completely screws you over.

It was gone quarter past ten now. Marc didn't know any more dormitory houses, and he was well away from the small area of Paris he knew well.

Rather than spend hours hunting for a room, with the constant risk of getting busted, he decided to bed down in the first decent hiding spot he found. Anything would do: a back garden, the landing on a set of fire stairs, or even a big dustbin.

The street Marc was in was mostly shops, with no gaps or hiding places between them. He glanced down side streets, but they all looked the same. When he got to the third street he had to dash, because he was within sight of a bar crammed with singing German soldiers.

The fourth turning looked more hopeful, but Marc only made ten paces past the corner when a torch beam lit him up. It shone from the doorway of a small Renault car with two gendarmes stepping out.

Marc thought about running, but the officer on the driver's side had his gun drawn, and you'd have to be a really bad shot to miss from this range.

'Get here,' the one with the torch said. He was in his forties, with the two middle fingers missing off his left hand. 'Do you have a night worker's pass?'

'No, sir.'

'Then we have a problem,' the officer said. 'Show me the rest of your documents.'

Marc nervously laid his documents on the roof of the car, where the officers inspected them by torchlight.

'What's your destination?'

'I thought I'd be able to get a room in a dormitory house before curfew,' Marc explained. 'But I've been to four and they're all full.'

'These documents are pristine,' the officer noted, as he flipped them over to look for the date-of-issue stamp. 'Issued today, in fact.'

The officer on the other side of the little car picked up Marc's employment status document. He was extremely tall, and the effect was exaggerated by his long coat and tall hat.

'You're from Paris?' he began.

'Yes, sir.'

'It says you are released from prison camp on compassionate grounds, to look after your family?'

Marc nodded. 'Yes, sir.'

'This form says you come from Paris, so why are you walking the streets looking for a dormitory house?'

'I . . .' Marc began, before coughing to buy thinking time. 'There's been a bit of a mix-up. I expect my older sister sent a letter saying where my family has moved to. But a prisoner's mail can take weeks to pass through the censors.'

'So how will you find your family?'

'I know where my aunt lives,' Marc said. 'But it's quite a way, near Beauvais. I didn't sleep on the train from Germany last night and I've been on my feet all day. It was getting near curfew and I just needed a place to sleep.'

To Marc's relief, the cops seemed to buy his story.

'If your aunt was in Paris, we might have been able to drive you,' the gendarme with the torch said, taking a friendlier tone. 'But Beauvais is over fifty kilometres, so you'll have to spend the night in a cell. In the morning you can take your train to Beauvais.'

'What's your aunt's address in Beauvais?' the tall cop asked.

'Fifteen Rue Lavande,' Marc said, picking a street name out of thin air, but hoping it would be OK if he sounded confident enough.

'Rue Lavande,' the tall officer repeated, as he tapped fingernails on the roof of the car. 'Hopefully you'll find your aunt there tomorrow, in the meantime get in the car. The station is only a two-minute ride.'

*

Every prison gets the same smell of bad food, bodies and cigarettes and Marc found it depressingly familiar. He was one of the first arrivals in a big holding cell that filled up as the night went on. There were several drunks, a husband spattered in his wife's blood, two Luftwaffe[8] pilots who'd fought over a girl and an old man who seemed completely off his head.

There was nothing but the concrete floor to sit on. Marc had barely slept the previous two nights and this one turned out no better, with constant noise and a crazy man who kept screaming and asking where all his teeth had gone.

There was light coming through the slot windows when the officer with the missing fingers called out

---

[8] Luftwaffe – the German air force.

Marcel DePaul. It took Marc a couple of seconds to remember that it was his new name.

'Sit,' the Gendarme said, when they reached a titchy interrogation room. All Marc's things were spread across the table. 'There's a train to Beauvais at 06:42, and they run every half-hour after that.'

'That's good,' Marc said guardedly. But he knew there was a sting in the tail because cops didn't take you to an interview room if they were just going to release you.

He'd only seen the gendarme the night before in the dark, and he looked more wrinkled in the morning sunlight.

'The only thing is, I called a colleague in Beauvais. He said he's lived there all his life, but he'd never heard of Rue Lavande.'

Marc rubbed his eye. He needed his wits, but a third consecutive night without sleep meant his brain was in fog. It was a struggle just to keep his eyes open.

'Maybe I got the name wrong. It doesn't matter. I know the streets. I'll find it when I get there.'

'Why don't you cut out the shit?' the officer shouted, as he banged his fist on the table. 'I'm trying to help, but you're treating me like a fool. You're carrying a brand-new set of identity papers but you're full of lies about family and places that don't seem to exist.'

'It's not lies,' was all Marc could muster, as he wondered whether the gendarme was a good guy who wanted to help, or a Nazi pet.

The problem was, interrogators are trained to put suspects off balance by switching from scaring the hell out of you to pretending to be your best pal, and that made it impossible for Marc to get any real sense of the gendarme's motivation.

'Carrying false identity papers can get you shot,' the gendarme said. 'I've been working all night and I need my beauty sleep, so here are two choices. One, give me a story that checks out, or two I'll pass your case on to a senior investigative team led by a German military officer. And believe me, you don't want that to happen.'

Marc realised he didn't have much choice but to trust his interrogator. If he was genuinely a good guy the gendarme might let him off the hook. If he wasn't, Marc was screwed no matter what he said.

And Marc did have one story that would check out. He locked his fingers together to stop his hands shaking and took a deep breath before speaking. 'My real name is Marc Kilgour,' he began. 'I ran away from an orphanage a few kilometres north of Beauvais, a while back.'

'This better check out,' the gendarme said suspiciously, as he wrote the information down.

He then asked for the exact address of the orphanage and the name of someone who'd be able to verify that he'd run away.

Marc thought of giving the name of Mr Tomas, the

orphanage director who'd beaten him on a regular basis, but Tomas was unpredictable so he gave the name of a young nun who'd always stuck up for her orphans.

'Sister Madeline knows all about me,' Marc said. 'And Sister Mary Magdalene, though she's getting on a bit.'

'I see,' the gendarme said. 'But these new identity papers are real. How did you obtain them?'

Marc was shattered, but only had seconds to think of a convincing way that a runaway orphan could get a full set of fake documents, without opening further lines of enquiry, or incriminating anyone else.

'When I was on the street this guy told me that returning prisoners can get identity documents if they present a travel warrant,' Marc said. 'I went through the rubbish bins outside the identity office. It took a few attempts, but eventually I found a travel warrant they'd thrown out. I flattened it out, changed the picture, and carefully altered some of the details before taking it back to the counter inside.'

The gendarme laughed. 'Ingenious,' he said. 'But *extremely* risky. And a serious offence in the eyes of our German masters.'

Having already thrown himself at the gendarme's mercy, Marc decided it was a good time to sound pitiful. 'I've never done anything like this before, sir. I was sick of the other boys, and I hate farm work.'

'Let me tell you,' the gendarme said, raising his voice

and wagging his pointing finger. 'An orphanage may not be great, but I've seen boys younger than you rot away in prison cells that would make your hair stand on end.'

'Yes, sir,' Marc said meekly.

'You're a very stupid boy,' the gendarme said, as he picked up Marc's papers and rattled them in his face. 'If you mess with the system, it will crush you.'

'I'm going out to call my colleague in Beauvais. He'll check out your story. Is there anything else you think you ought to tell me before I go?'

'Not that I can think of, sir.'

*

A two-hour wait in the interrogation room did Marc's nerves no good. An orderly brought him stewed oats for breakfast, but they stuck in his throat.

'Mr Kilgour,' the gendarme said warmly, when he finally got back. 'My colleague in Beauvais actually knows your friend Sister Madeline. He rode out to the orphanage and the sister vouched for you. He even dug up a missing person's report, filed by the orphanage director.'

Marc smiled, feeling slightly embarrassed for some reason. 'So, what happens now?'

'One of my fellow officers will escort you back to Beauvais,' the gendarme said, before leaning across and squishing Marc's cheeks in a manner that was friendly, but also intimidating. 'You're *bloody* lucky that I picked you up, instead of a German patrol.'

'I'm sure I could find my own way,' Marc said.

The gendarme flashed with anger. 'To go on the run again?'

'I just don't want to waste your time, that's all,' Marc said.

'Besides, you can't travel unaccompanied. You've got no papers, have you?'

Marc knew better than to suggest returning the false ones.

'I'll have a report typed up, saying that you've reported your identity documents missing,' the gendarme said. 'The orphanage shouldn't have much problem getting you new papers if they have that.'

CHAPTER NINETEEN

Marc stared at the shabby three-storey orphanage like a ghost raised from the dead. He'd lived here for the first twelve years of his life, one of a hundred orphan boys looked after by nuns, under the command of Director Tomas and his steel-tipped cane.

Marc's departure two years earlier had involved stealing clothes and boots from fellow orphans, smashing a kid's head through a pane of glass and robbing Director Tomas' bicycle. But while Marc didn't expect an easy ride either from staff or the other kids, the orphanage offered food and shelter while he built up his strength and planned the next stage of his escape.

And yet sisters Madeline, Mary and Raphael ran smiling when they saw Marc approach with the gendarme.

'God be praised,' Madeline gushed, as she pulled Marc

into a tight hug. 'You were in our prayers every day.'

In the two years Marc had been away it had never once occurred that he'd been missed by the women who'd fed him, scrubbed his clothes and cared for him since he was a baby. The orphans tended to think of the nuns as harsh and slightly witchlike, but Marc now had enough of an outsider's perspective to see that a hundred boys took a lot of looking after. It wasn't that there was no love in this place, just that necessity meant it was spread thin.

Little details filled Marc with nostalgia. The creak of the main door, the way morning sun reflected off the highly polished floor. Above all, the smell of boys – socks, bad breath, pee and farts. It wasn't nice, but it had an innocent charm when compared with the stench of BO and cigarettes made by grown men.

Sister Mary looked at the elderly gendarme who'd accompanied Marc from Paris. 'Would you like a plate, sir?'

They went to the kitchen. Marc got bread, cheese and carrot, while the gendarme was also treated to wine and chopped sausage.

The nuns spent their lives boiling clothes, bleaching toilets and scrubbing floors with scalding water. As Marc ate, he found Sister Madeline hovering over his shoulders, inspecting him for lice. His stubbly hair passed muster, but she made a tiny yelp when she pulled back his shirt collar and saw a pair of tiny body lice.

'I'll draw a hot bath,' she said urgently. 'He needs a *thorough* delousing and we can soon boil his clothes.'

After a second glass of wine, the gendarme shook Marc's hand and told him to look after himself. Marc thanked him, and repeated his promise to behave.

The two things that had most struck Marc since arriving back in France were the vast resources the Germans had to devote to keeping the French population in line, and the fact that officials such as gendarmes and the woman at the identity office weren't keen on enforcing German regulations. Even the girl who'd been snogging a German had sided with him.

For all the propaganda posters, smart German uniforms and swastikas flying from important buildings, Marc got the feeling that occupied France was like a beautifully iced cake, being devoured from within by an army of tiny pissed-off ants.

Marc went rigid when he recognised the distinctive click of Director Tomas' office door. But he glanced back into the hallway, and was surprised to see Sister Raphael step out of his office. She was a short woman, with pudgy white hands that always reminded Marc of balls of dough.

'Has Director Tomas said anything about me?' Marc asked nervously.

Sister Raphael smiled. 'The director is no longer in the

service of the church. We sisters run the orphanage now, under direct authority from the bishop.'

Marc didn't know what to say, but his relieved smile said enough.

'We seem to get along well enough without his guidance,' Sister Raphael said, a cheeky smile creeping into her stern expression. 'But just because you're taller than me, *don't* think I can't take a cane to your backside if you err from the path of righteousness.'

Marc had never heard the nuns openly criticise Director Tomas, but they'd never liked the brutal way he'd treated the orphans and Sister Raphael was clearly well shot of him.

'Is Tomas dead?' Marc asked hopefully.

'The Germans took a liking to our director,' the sister explained. 'He's working for the Requisition Authority in Beauvais. But he still comes back to his cottage at night, and you'd do well to steer clear of him. He wasn't very happy when you stole his bicycle.'

'Avoiding him sounds sensible,' Marc agreed.

'I'll need to get a photograph for your identity papers,' Sister Raphael said. 'I'll tell them you were born in 1929, instead of 1928, which will keep you out of the factories for another year. You used to work for Morel the farmer, didn't you?'

Marc nodded.

'I'll send you to work for him. No point restarting

school at your age after two years absent.'

Marc's mouth twisted awkwardly. 'The last time I saw Morel, he threatened to chop my . . .'

Marc tailed off, because he didn't know how to refer to his balls in front of a nun.

'Farmer Morel needs labourers,' Sister Raphael said. 'And this orphanage relies on the extra food he supplies. He'll work you hard, your wages will be used to pay for the boots you stole and the glass you broke when you ran away.'

'I *kind of* tripped Farmer Morel's daughter into a pit of manure,' Marc said delicately.

Sister Raphael wagged her finger. 'He's a good Christian. I'm sure he'll forgive you.'

Marc wasn't convinced, but didn't argue because in the twelve years he'd lived in the orphanage he'd had hundreds of disputes with the nuns and had won precisely none of them.

'Sister Madeline will have the water hot for your bath shortly,' Sister Raphael said. 'Go help her drag the tub out into the yard.'

*

Even in freezing weather, orphans bathed in a tin bath on the patio behind the kitchen. On busy days, three baths were lined up, with boys hopping in one after another and the nuns scooting back and forth from the kitchen with hot kettles to warm up increasingly filthy water.

Sister Madeline was in her twenties, and Marc suspected she was rather sexy behind her black habit and veil. She'd probably seen him naked a hundred times, but puberty had properly kicked in and he turned bright red as he stripped naked. He dipped a toe in the steaming water, but that only earned a rebuke.

'Not yet! I've got to get the delousing powder.'

The white powder came in a huge aluminium can. Sister Madeline used an enamel mug to scoop some out. She sprinkled it over Marc's head, then followed up with a bucket of steaming hot water.

'Oww, Jesus!' Marc moaned.

'You'll burn in hell taking our lord's name in vain,' Sister Madeline snapped furiously, as she gave his wet back a hard slap. 'And fancy a big boy like you making more fuss than a six-year-old!'

The insecticide powder fizzed on contact with water, and Madeline tipped another cupful into Marc's hands and told him to rub it into all his hairy bits. When he was white all over, the nun left him to cook naked in the sun, while she sprinkled more delousing powder over a smaller tub filled with boiling water into which she dunked Marc's clothes.

The older orphans were out at school, but Marc being white and naked greatly amused a group of three- to six-year-olds who'd emerged from a game in the surrounding fields with a teenage nun who Marc had never seen before.

'He ran away,' one of the oldest boys said authoritatively. 'He's called Marc.'

After ten minutes standing with the fizzing insecticide powder on his skin, Sister Madeline reduced the young spectators to howls of laughter as Marc hopped about under a cold hose, before finally allowing him to settle in the tin bath.

There was no soap and Sister Madeline insisted that Marc scrubbed himself thoroughly with a pumice stone, before she broke off and mischievously chased the little spectators with the hose. Marc sat in his bath with his head tipped back, watching the little boys throwing off their shirts and squealing under the cold water.

If Director Tomas was still around, he'd have stormed out of his office, grabbed the first two noisy kids he could lay hands on and brought them inside for a good thrashing. The orphanage was a happier place for his absence, and Marc was glad to know it.

For two years, he'd seen brutality and suffering, as the Nazis turned Europe into a gigantic slave camp. But here, in a little orphanage in the back of beyond, he'd found something that had actually got better.

*

It was only 12:30, but the nuns had seen how skinny Marc had got and they let him eat again when the six- to twelve-year-olds arrived back from the village school and ate lunch.

'AAARGH!' Jacques screamed, fighting off happy tears as he gave Marc a hug. 'I can't believe you're back.'

Jacques was eleven now. He'd slept in a bunk below Marc from his fifth birthday when he'd been moved out of the nursery, until the day Marc ran away. He was the nearest thing Marc ever had to a little brother.

'You've really grown,' Marc said, as he noticed a surprising lack of familiar faces among the boys returning from school. Clearly the war had made an excellent job of creating new orphans.

'You remember Victor?' Jacques asked. 'He got moved up to our attic dorm. We're mates now.'

'I remember trying to steal a pair of boots, and you trying to stop me with a broken arm,' Marc said, putting the truth right out in the open and hoping Victor wasn't holding any grudge.

'Everyone thought you was dead,' Victor said.

'Nope,' Jacques said. 'I *never* believed that.'

'You bullshitter,' Victor scoffed. '*Everyone* thought Marc was dead. I mean, no other kid ever ran away for more than about two weeks.'

'OK, I had my doubts,' Jacques admitted. 'But I always *hoped* you were alive. So where have you been? What have you been doing?'

CHAPTER TWENTY

Marc couldn't tell his fellow orphans the truth about the last two years of his life. Even if he had, none would believe that he'd got mixed up in plans to steal the blueprints of a miniature radio transmitter, helped destroy Nazi plans to invade Britain, then escaped across the Channel, where he'd been given espionage training by the British Secret Service, before being captured and sent to prison in Germany while on another undercover mission to sabotage U-boats.

When Marc ran away he'd never been further than Beauvais, six kilometres south of the orphanage. He'd never been on a train, never eaten in a cafe or restaurant. That was still true for most of the kids in the orphanage, so they were easily impressed.

After a satisfactory afternoon nap on clean orphanage

sheets, Marc woke just before 6 p.m. and found a group of little kids badgering him with questions. It was all little stuff, like what the Paris Metro was like, had he seen a Panzer tank, was Paris near the sea?

The older boys were more interested in how Marc had survived on his own for two years. He told them he'd spent the whole time in Paris, living in empty houses, making money doing odd jobs. He even jazzed the story up with a couple of hot girlfriends.

Marc got full-on hero worship from everyone until his old nemesis Lanier returned from his job at the local bakery.

There's usually a pecking order with kids: tough dominates weak, clever outsmarts stupid, tall looks down on short. Marc and Lanier's mutual hatred grew out of the fact that they were so alike. Same age, same build, both quite clever. Lanier probably had more of a nasty streak, but in the charged and occasionally brutal atmosphere of a boys' orphanage Marc had been no angel.

As they sat at tables behind the dilapidated orphanage eating their evening meal – the first time Marc had eaten three meals in a day for what felt like about a million years – Lanier squatted on the table's edge in a sweat-stained baker's overall and took every opportunity to remind the others that during Marc's escape, he'd stolen a boy called Noel's working boots and smashed another boy's head through a window.

'Sebastian's got scars on his cheek,' Lanier said. 'He's working in Germany now, but you'd better steer clear if he ever comes back.'

There used to be kids of sixteen and seventeen in the orphanage, but Marc had noticed there was nobody older than fourteen now.

'Are they all in Germany?' Marc asked, depressed at the thought.

Forced labourers got paid wages and were treated better than prisoners, but there wasn't much in it.

'Some work in factories here,' Jacques explained.

'Noel?' Marc asked.

'Director Tomas sent him to work on the wall, I think,' Jacques said.

'Wall?' Marc asked.

'Atlantic wall, dummy,' Lanier said, delighting in Marc's ignorance. 'Building defences along the coast, in case the Yanks and Brits invade.'

'Ahh,' Marc said.

He'd rarely heard the news in prison camp. When you did it was hard to separate facts from rumours, but the tide of war did appear to be turning: twenty months earlier, Marc had helped to destroy barges for a planned German invasion of Britain. Now, Hitler was building coastal defences to stop Britain and America coming in the opposite direction.

'When are they expecting the invasion?' Marc asked.

'Everyone says it has to be summer to invade,' Victor said. 'Didn't you see any news in Paris?'

Marc realised he'd slipped up. If he'd really been in Paris for two years, he'd know a lot more than a bunch of kids in a remote orphanage.

'I was busy looking after myself most of the time,' Marc said unconvincingly.

'I reckon they'll invade tomorrow afternoon,' Jacques joked. 'Liberation by Sunday teatime.'

Marc was torn: he liked the idea of the Nazis getting their arses kicked, but was scared by the prospect of a brutal land war. The German army had swept into France with a rapid and relatively bloodless invasion two summers earlier, but he doubted they'd surrender anything like as easily.

'And did you say Director Tomas sent Noel to Germany?' Marc asked.

Jacques nodded, as a couple of little kids who'd grown bored peeled off to play.

'Tomas loves the Nazis,' Lanier said, for once turning his bitterness on to something other than Marc. 'He's in charge of the Requisition Authority.'

'What's the Requisition Authority?' Marc asked.

Lanier shook his head, as if Marc was stupid for not knowing. 'If the Germans want something – cattle, food, workers, ammunition, you name it – it's the Requisition

Authority's job to get hold of it.'

'Last time he came to the orphanage, he started shouting at the nuns and left with four boys,' Jacques explained. 'Lanier was lucky – he was out working.'

'Bastard,' Marc said bitterly. 'He always loved having any kind of authority.'

'I've heard it's not bad in Germany,' a lad Marc didn't know said. 'They get paid in Reichsmarks, and I'd rather work in a factory than on Morel's stinking farm.'

Marc would have loved to set the kid straight, but couldn't say anything about Germany without blowing his cover.

All the boys looked round as the teenage sister, who was called Peter, came out into the yard clapping her hands.

'Right,' she yelled. 'Who's been leaving their empty pots on the tables and going off to play? You, you, you, you, get inside and help with the washing-up. Lanier, Jacques, get mops. The stairs leading up to the dormitories are a disgrace.'

Lanier pointed at Marc, 'What about His Majesty? I've been working since six this morning. What's he done but sit on his skinny butt all day?'

'Language,' Sister Peter said firmly. She might have been cute, but she must have been strict because all the kids did what she asked without any messing. 'Marc,

Sister Raphael says you need to go across to Morel's farm. Tell him you're a reformed character and ask for your old job back.'

Not all the boys knew the story about Marc tipping Morel's daughter Jae into a slurry pit, but those that did looked aghast.

'Be *reasonable*,' Marc said. 'I'm really tired. Can't it wait 'til tomorrow?'

To Lanier's delight, Sister Peter pivoted on her skinny legs and threatened Marc with a slap.

'No nonsense,' she said. 'And you haven't got any documents yet, so stay off the roads.'

*

The Morel family had the grandest house in the area, three storeys high with a large stable block off to one side. The path up to the house was daunting, and a long wait in a double-height hallway, with oil paintings and a ticking grandfather clock, frayed Marc's nerves.

Farmer Morel was rich and influential, but there was nothing about him that would make you think him special, in his rough peasant shirt and wooden clogs.

'Kilgour,' Morel began, with the calm authority of a man who was used to people jumping when he told them to. 'Was *something* not clear when I told you never to set foot on my land again?'

He'd just finished his evening meal and was pulling a napkin out of his shirt. To Marc's surprise, a glimpse into the dining room showed that he'd been eating with two Luftwaffe officers.

'Sister Raphael said you were short of farm hands. I told her you wouldn't want me, but she asked you to reconsider as a personal favour to her.'

Morel's eyes narrowed at the mention of Sister Raphael. 'I do owe her a favour,' he said grudgingly. 'And I never found fault with your work, so be at my farm manager's office, six sharp tomorrow morning. And be prepared to work *hard*.'

'I'm a hard worker, sir,' Marc said, doing a good job of hiding his disappointment. He'd have been content to doss around the orphanage for a few days while Sister Raphael tried finding him another job.

Morel leaned forward and jabbed the ill-fitting shirt that Sister Madeline had found for Marc to wear while his clothes dried.

'But here's fair warning,' Morel growled. 'Don't speak, don't go near, don't even *look* at my daughter. If you do, I'll have a couple of my biggest farm hands strap you to a barn door, and horse whip you. Is that clear?'

'Hard to see how it could be any clearer, sir,' Marc said warily.

'Right, I'm off for my dessert. Don't be late in the morning.'

Marc felt like he'd come full circle: back at the orphanage, back working for Morel. As he crossed the field back towards the orphanage, another piece of his past scrambled through shoulder-height corn towards him.

Jae Morel was fourteen. Marc had had a crush on her since he was eight years old, but now Marc had hormones and Jae had tits. The mix of the two was like a kick in the gut.

'Marc.' Jae grinned. 'It's you! When did you get back?'

Marc glanced about and backed up.

'What's the matter?' Jae asked, as she stepped out of the corn.

Jae had always been a posh girl: pigtails and violin lessons. But this 1942 model Jae was different: mucky rubber farm boots, blue farmer's smock, scarf over her hair and sweat streaking down a dirty face.

'Being near you could seriously endanger my health,' Marc said seriously. 'I met your dad. Horse whipping was mentioned.'

'I shouldn't have run to my dad that day,' Jae said, looking down at her boots in shame. 'I knew exactly what Director Tomas would do to you, but I was a snotty little so-and-so back then. Look at me now though.'

Jae held out her hands, showing dirty fingers and cracked nails.

'My school closed down,' Jae explained. 'Most of the other rich kids went south during the invasion and never came back. My dad's desperate for labourers, so here I am, up to my elbows in dung every day.'

Marc smiled at Farmer Morel's muddy little princess, but kept looking over his shoulder nervously.

Jae scoffed. 'You're really *that* scared of my dad?'

'Everyone's scared of your dad. I was just at your house. He was eating with Nazis!'

'Luftwaffe headquarters is in Beauvais,' Jae explained. 'We've got a big house, so they've billeted four senior officers with us.'

'So he gave me a job,' Marc said, as he kept looking around.

'He's angry about you tossing me in that manure pit.'

'Tell me something I *don't* know. And in case you don't remember, we were arguing and you slipped into the pit by accident.'

Jae shrugged. 'I was sad when you ran away, Marc. I always liked you.'

'I like you,' Marc said. 'But you've got to stay away from me.'

'When you were in first grade you never had any shoes,' Jae said. 'Your feet were filthy and the teacher went mad because you sat at your desk with your foot in your mouth chewing your big toenail.'

Marc laughed. 'Oh my god! I'd *totally* forgotten that I used to be able to do that.'

'Maybe you still can,' Jae said.

'I'll give it a try,' Marc said, laughing uncomfortably.

'Look, Jae, I've got nothing against you. I've just got no desire to get strapped to a barn door and horse whipped.'

'You're cute when you're scared,' Jae said, as she stepped up to Marc and kissed him softly on the lips.

Part Two

August–September 1942

CHAPTER TWENTY-ONE

Wednesday 19 August 1942

Seven weeks after arriving at the orphanage, Marc had a modest growth of hair on his head and the nuns had fed him back to a healthy weight.

Although Farmer Morel's main crop was wheat, he also kept cows and chickens, bred working horses, ran a small orchard and had a twenty-acre vegetable patch where Marc had spent the morning pulling carrots out of the ground.

Marc took his lunch of bread, cheese and freshly squeezed milk while lying in tall grass with an unbuttoned shirt exposing his chest to the hot sun.

'I sneaked out a jar of honey,' Jae said, when she sat down alongside him.

Her boots were hurting and she pulled them off,

revealing narrow sockless feet, beaded with sweat.

'Did you ask your dad for new boots?' Marc asked, as he pulled Jae's foot into his lap and started gently massaging her toes.

'They stink,' Jae said, as she gave Marc a kiss on the cheek. 'My dad says you can't get farm boots anywhere. But one of the Luftwaffe officers staying at our house said he's trying to get me a pair of flying boots.'

'Is that the creepy one with glasses?'

'Jealous?' Jae laughed, as she opened a little cloth sack and unscrewed the lid from a jar of honey.

'If you want to flaunt yourself to some pot-bellied Nazi that's your lookout,' Marc said. 'But I bet he wouldn't give you nice foot massages like me.'

To make his point, Marc lifted Jae's leg up and kissed the ball of her foot. Then he coughed.

'Actually, they *really* stink,' he croaked.

Jae slumped back in the grass and stated to laugh. 'What did you expect after six hours' farm work, a rose garden?'

Now Jae was flat on her back, Marc rolled over and tried to kiss her. But before he got close, Jae reached up and pushed a spoonful of honey into his mouth.

'Mmm.'

Jae ate a spoonful herself and they started kissing, with the honey gluing lips and tongues together. As she reached up to put her hand on Marc's bum, Marc slid a hand under the strap holding up her overall and cupped her breast.

'I think about you all the time,' Marc said when he came up for air.

'I looked on the work rota in the manager's office,' Jae said. 'We're both off Friday. You fancy swimming down by the lake? I'll sneak into our kitchen after the maid goes to bed and make a picnic.'

Jae was turning Marc on like crazy, but his face gave away his uncertainty.

'Coward,' Jae said. 'Sneaking around is half the fun.'

Marc laughed. 'If we're caught you'll get told off by Daddy. *I'll* get thrashed.'

'Aren't I worth it?' Jae asked.

'Until something better comes along,' Marc teased.

Jae gave Marc a dig in the ribs. As he laughed and rolled away clutching his side, he spotted the elderly farm manager coming across the field towards them.

'It's Felix,' Marc blurted.

Jae grabbed her boots and lunch bag, and gave Marc a quick peck on the cheek before scuttling behind the nearest barn.

'See you later,' she whispered.

Marc stood up as Felix approached. Like any boss there were times when Felix complained about something Marc had done, or forced Marc to do something he didn't want to, but on the whole he was decent as long as you didn't muck about.

'Dangerous game you're playing,' he said, half smiling as Marc got out of the grass.

'You won't tell Morel, will you?' Marc asked.

'She's not my daughter,' Felix laughed. 'All I care about is your work, and you do that well enough. Do you think you could manage taking the delivery cart into Beauvais this afternoon?'

'I've run a delivery cart before,' Marc said, nodding. 'But what about Nichol?'

'His little one's sick. All the deliveries are written up in the ledger. I'll get a couple of the younger orphans to finish off your carrots when they get here after school.'

*

Rationing, poverty and the absence of men meant cafes and bars had fared badly in most small French towns, but Beauvais had become a Luftwaffe town, with combat airfields for fighter and bomber squadrons nearby, regional headquarters in a chateau on the town's edge, and a busy cargo and passenger airport at Tille, two kilometres north-west. The air war ran day and night, so you were as likely to find a crowded bar on Beauvais' main drag at seven in the morning as seven at night.

While most of Occupied France was controlled by the German army and Gestapo, Beauvais and its surroundings were a special military zone, under command of the Luftwaffe. In contrast to Marc's experiences in Paris and Lorient, the Luftwaffe police took a relaxed attitude. They usually left routine law enforcement to local gendarmes, so he was surprised to get held up in a long queue at a snap checkpoint on the edge of town.

The Luftwaffe officer only gave Marc's identity documents a brief glance and couldn't have cared less about the supply of black-market food on the back of his cart.

'Have you seen any sign of paratroopers?' the young officer asked gravely.

'No,' Marc replied.

'Any other unusual movements? Trucks? Men or vehicles on the back roads? Equipment drops?'

'Nothing,' Marc said.

'What have you been doing today?'

'I work on Morel's farm a few kilometres north,' Marc said. 'I spent the morning picking carrots.'

'Do you have family?'

'I live in an orphanage.'

The officer seemed pleased to hear this. 'When you get home tonight, tell all the boys there's a fifty franc bounty for any lad who unearths suspicious activity or Allied soldiers.'

The Luftwaffe facilities around Beauvais were high-value sabotage targets and Marc wondered if there had been a resistance operation in the area.

'Are you looking out for anything in particular?' he asked.

The officer shrugged. 'We've not been given details, but we've been put on our highest alert status. All our leave has been cancelled and there's an emergency 8 p.m. curfew tonight, for French *and* off-duty Germans.'

The extra security made Marc curious, but there was no panic in central Beauvais. Relaxed figures in Luftwaffe uniforms dominated the bars and cafes as Marc delivered vegetables and bags of flour. He'd never driven the delivery cart before and everyone kept asking if Nichol was OK.

Marc had only been to Beauvais a couple of times in the seven weeks since he'd returned and didn't know anyone well enough to ask what they knew about the curfew and the extra security. But he did overhear tantalising snippets of conversation between two senior Luftwaffe officers drinking wine in a cafe.

They didn't suspect that the French boy in grubby farm overalls understood German and while the background noise made it tough to catch every word, he picked up several references to a large-scale Allied landing in Dieppe that morning.

The two horses were sweating from the heat, so Marc stopped at the next drinking trough, doused them in cool water and gave them a handful of bushy green carrot tops.

'Well, well,' Mr Tomas said, as he stepped across the cobbles. 'Look who's back in town.'

Marc baulked as he recognised the face from all his childhood nightmares. This was the former orphanage director who'd thrashed him on at least a hundred occasions, put him on bread and water for talking back and made him spend a week sleeping in a frozen barn for walking upstairs too noisily.

Tomas was physically strong, but Marc drew some confidence from the fact that he was now as tall as his former tormenter.

'Hello, sir,' Marc said, feeling like he was five years old again.

Tomas wore black trousers and a brown shirt with a red swastika armband. It reminded Marc of Sivertsen, the Danish prison guard aboard *Oper* who'd always been keen to prove that he was a better Nazi than any German.

'Where are these vegetables being delivered to?' Tomas asked, as he lifted up a couple of boxes to see what was inside. 'Do you have proper documentation?'

Marc wondered if he was supposed to be paying Tomas some sort of bribe. 'I took over the delivery route for one day,' Marc said. 'I don't know all the details, but I'm sure Farmer Morel keeps things in order.'

'Ahh, so this is Farmer Morel's cart?' Tomas said, puffing his chest out importantly. '*He* ought to know that all farm produce must be registered with and sold through the Requisition Authority.'

'I'm just a delivery boy,' Marc said irritably. 'Maybe you should speak to Morel about it.'

Tomas raised an eyebrow. 'And what about my bike? I reported it stolen, you know.'

'It never got far,' Marc said. 'A tramp ran off with it before I even got past Beauvais.'

'I'm sure the gendarmes would be most interested to know that you're back in town, Kilgour. Stealing a bike is a serious offence.'

Marc would have loved to punch his former tormentor in the mouth. But while Tomas had aged in the two years since Marc last saw him, this was still a man who'd regularly kicked well-built sixteen- and seventeen-year-olds around the courtyard behind the orphanage.

'I've got a job to do,' Marc said. 'If you want to know about the cart, speak to Morel about it.'

'I've got my eye on Morel,' Tomas said, cracking the exact smile that used to come before a beating. 'If you keep me well informed every time food leaves Morel's land without clearance from the Requisition Authority, I might forget to remind the local gendarmes that you stole my bike.'

Marc snorted. 'I'd just deny it,' he said. 'Nobody saw me. There were bombs going off, and thousands of troops in retreat that day. You'd never prove a thing.'

Tomas reared up on his heels and hissed, but Marc was determined to show that he wasn't scared of his old tormentor.

'Look at you with your *stupid* little trousers and homemade swastika,' he spat. 'When this war's over, traitors like you will be put up against a wall and shot.'

CHAPTER TWENTY-TWO

When Marc got the cart back to the stables beside Farmer Morel's house he thought he ought to warn someone about his encounter with Director Tomas. Felix the farm manager wasn't in his office, but Marc bumped into Morel and told him what had happened instead.

'Tomas has always had it in for me,' Marc explained. 'I hope I didn't do anything that backfires on you.'

Marc felt anxious as Morel looked him up and down.

'Felix says you're a good worker,' Morel began. 'A quick learner, who can be left alone to get on with things. And it must have taken some intelligence, surviving on your own as a runaway these past two years.'

'It wasn't easy, sir,' Marc answered awkwardly.

'Step inside,' Morel said. 'I'll show you what Felix and I have to put up with.'

Morel led Marc back into the farm office, which was actually two stables knocked into one. There were bundles of unbound paperwork stacked from floor to ceiling.

'The Germans make me account for everything I grow, down to the last grain of wheat,' Morel explained. 'The Requisition Laws say I have to sell all the food I produce to the Requisition Authority, which is run by your old friend Mr Tomas. Even if I want to use grain I've grown to feed my own livestock, I'm supposed to sell it to the Requisition Authority and buy it back from them at three times what they pay me.'

'That's mad,' Marc said.

Morel nodded. 'And the prices the Requisition Authority buys my food at are below my cost of production. So every farmer *has* to sell locally on the black market to survive. But I can be given a lengthy prison sentence for selling black-market food. The system effectively turns every farmer in France into a criminal.'

'That makes no sense,' Marc said.

'Doesn't it?' Morel asked. 'The Germans like having us all walking on eggshells. They've got half a dozen reasons to lock me up if I do anything that displeases them. It doesn't help me to sleep at night.'

Morel had always struck Marc as a towering, all-powerful figure, so this vision of him as a pawn trapped between conflicting Nazi demands was an eye-opener.

'Your old friend Mr Tomas has had his Requisition Authority inspectors up here half a dozen times: poring over ledgers, hunting through barns and counting cauliflowers,' Morel explained. 'One time he even had all my manure weighed, to see if I was hiding cows.'

'Sounds stressful,' Marc said. 'Makes me glad I'm a nobody.'

Morel laughed noisily. 'Fortunately, I'm friendly with a couple of the senior officers billeted at my house,' he explained. 'The Luftwaffe are keen for the cafes they frequent in Beauvais to serve decent food, so some nice cheese and the odd bottle of wine dropped into the right Luftwaffe officer's pocket helps keep Tomas off my back.'

'I got the impression that he had it in for you,' Marc said.

'Tomas fell out with me when he was orphanage director,' Morel explained. 'When my regular labourers got conscripted into the army before the invasion, he charged me an absolute fortune to employ boys like you from the orphanage. By all accounts Tomas was keeping some of the money for himself. When the Bishop found out, Tomas got sacked. He thought I'd spoken to the Bishop, but I had no idea what was going on.'

Marc nodded. 'I'd bet it was one of the younger nuns. They would never say anything to Tomas's face, but they hated the way he treated us orphans. It wasn't just older kids like me. He'd use his cane full force on little five-

and six-year-olds. You'd see them stagger out of his office, clothes soaked in blood.'

'And now he parades around town wearing his little swastika, poking his nose into every farmer's business,' Morel said. 'So keep your eyes open and be sure to let me know if you hear anything else about him.'

'For sure,' Marc said. 'I've got as much reason to hate Tomas as you have.'

As Marc followed Morel out of the office, Jae was crossing the courtyard outside. Marc suspected she'd been listening.

'Daddy,' Jae said brightly, as she gave her father a kiss on the cheek. 'I was looking for you all over. There's a rumour going around that the invasion has started. Allied troops landed in Dieppe this morning.'

Marc felt awkward in the combined presence of the girl he was crazy about, and the father who had threatened to thrash him if he so much as looked at her.

'I'd better head back to the orphanage,' Marc said, shuffling past stacked papers towards the arched doorway.

'Did you hear anything about an invasion in town?' Morel asked.

'There was extra security on the main road,' Marc replied. 'They asked me if I'd seen any paratroopers. And I overheard some Germans saying that all their leave had been cancelled.'

'Must be some truth in it then,' Morel said thoughtfully.

Jae broke into a huge grin. 'Daddy, this could be it,' she squealed. 'The Allies could really be coming! We could be free again.'

Morel furrowed his brow and stroked the hair on his chin. 'Or we could find ourselves stuck in the middle of a long, bloody battle.'

'Might be an exercise,' Marc suggested. 'You know, making sure all the Germans are on their toes.'

'That would make sense,' Morel said. 'Hopefully we can listen to the BBC tonight, if they're not jamming the signal.'

Jae nodded. 'You know, Daddy, I've hardly got any friends my own age now. Marc's been working really hard on the farm. Do you think he could have dinner with us tonight?'

A look of horror flashed across Morel's face and Marc's heart rate shot up, but Jae kept pleading eyes fixed on her daddy.

'Well, I suppose,' Morel said grumpily. 'Tell cook to expect an extra guest. And you'll have to find him some decent clothes if he's going to eat dinner with us.'

*

As a dinner guest Marc could have used the front door of the Morels' grand house, but the habit of a lifetime meant he used the servants' entrance around the back. A young servant girl led him up to the main hallway, where Jae spotted him straight away and led him up to the first floor.

'Are you mad?' Marc asked, as he walked behind Jae. 'Why did you ask if I could come to dinner?'

'Food's nice here,' Jae said airily. 'Daddy needs a chance to get to know you. And you were getting on well enough in the office earlier.'

Marc found himself being led into a young man's bedroom – it belonged to one of Jae's older brothers, who'd been captured during the invasion and had spent over two years held prisoner in Germany.

'Put that on,' Jae said, as she pulled a tie out of a wardrobe.

'How?' Marc asked. 'I've never worn a tie in my life.'

'Ignoramus,' Jae said, half joking.

She kissed Marc quickly before pulling up his collar and starting to knot the tie.

'Daddy thinks you're a good worker. I've told him you're my friend. He misses having my brothers around, so just laugh at his jokes and pretend that you're interested in car engines.'

Marc didn't seem sure. 'Inviting me to dinner and making me put on your brother's tie won't cut it. Orphans don't mix with posh little girls like you.'

'Posh little girls don't shovel manure on farms either,' Jae said, as she pulled up the tie and turned Marc towards a mirrored wardrobe door. 'You don't scrub up too badly. And those boots are very nicely made. Where did they come from?'

Jae had only heard Marc's cover story about spending two years in Paris, so he couldn't say he'd stolen them from a dead soldier's bedroom in Germany.

'Somewhere down the line,' Marc said vaguely.

Farmer Morel was already at the dining table when they got downstairs. Jae's mother had died when she was seven, and her father had remarried a woman called Stephanie who spoke so rarely that she barely seemed to exist.

The final guest at the table was Karsten, one of the four Germans who lodged at the Morel house. He was a senior Luftwaffe pilot, but out of respect for his French hosts he'd changed out of uniform into corduroy trousers and a hand-knitted cardigan that gave him the look of a teacher or university professor.

With several hundred acres of land under their management, no member of the Morel family ever starved. Marc was impressed by a starter of baked Camembert cheese, served with apricots, grapes and the first white rolls he'd tasted since leaving England over a year earlier.

'So, Herr Karsten, should we be concerned at the security alert in Beauvais today?' Morel asked, as he wiped a generous lump of butter on to his roll.

Karsten gave a conspiratorial smile, the kind that says *I shouldn't talk about this, but I'm going to anyway*. 'The information we heard was patchy.'

'Was it an exercise?' Morel asked.

'No, I believe there really has been an Allied landing at Dieppe. Bigger than anything that's happened before, but nothing like a full-scale invasion. The oddest thing is that the Allies on the beach had almost no air support.'

'We *must* listen to the news on the wireless,' Jae said, as she pushed her foot towards Marc beneath the dining table and stroked his calf with her toes.

Marc almost inhaled a grape. And while the foot felt nice, he didn't like the way Jae regarded her father's threats as a big joke.

'Perhaps we should see what the BBC says in their seven o'clock bulletin,' Karsten said.

Marc looked around to see what the reaction to this was: most French people listened to British radio broadcasts, but it was illegal, so Marc wasn't sure if the German's suggestion was serious.

Apparently it was, because after a main course of beef and potatoes cooked in red wine gravy, Jae went off to warm up the valves of the radio in the drawing room and the cook was asked to bring coffee and desert through on a trolley.

For all Farmer Morel's difficulties with the German occupation, his family diet certainly wasn't suffering. As

Jae fiddled with the radio, Marc sat in a winged armchair, thinking guiltily about hungry prisoners as he ate strawberries and whipped cream, served on a bed of warm puff pastry.

BBC France fought a constant battle with German attempts to jam its signal, but after much fiddling Jae managed to tune a voice that was echoey but mostly comprehensible:

The ministry of war has announced . . . Small-scale raiding force landing at Dieppe, beginning at 5:50 this morning . . . Three thousand Canadian and British troops landed . . . Critical objectives were achieved, but the force withdrew after heavier than expected casualties. The exact objective of the raid remains classified, but it is believed to be the beginning of a summer campaign in which German defences around occupied France are probed by . . .

The radio broadcast broke up for several seconds after that, and when it came back the news had moved on to a story about American military production: *eight ships, fifty bombers, ninety fighters, two hundred tanks and two thousand freshly trained soldiers every single day. All lined up and ready to crush the Hun!*

'Not full-scale invasion, as I suspected,' Karsten said.

The German was clearly relieved, but also aware that this might not be regarded as good news by the people in whose home he was staying.

'I wonder if there will be a full-scale invasion attempt

this summer,' Jae's stepmother said.

'The Americans are nowhere near ready,' Karsten said. 'U-boats are playing havoc with their shipping, and autumn is almost here.'

Farmer Morel nodded. 'But with so much effort going into the Atlantic Wall, Britain and America will surely want to invade across the Channel as soon as it's feasible?'

'I'm a pilot, not a field marshall,' Karsten said. 'But if I had to bet on an invasion, I would guess June 1943. But if Russia is defeated first, a great mass of forces can be moved from the eastern front and an invasion would surely be impossible.'

Marc was intrigued by Morel and Karsten's frank discussion of the war, but with no imminent invasion, Jae had lost interest and tugged on his arm.

'Permission to be excused, Daddy,' Jae said politely. 'May I take Marc up to my room and play some records?'

Marc already felt that Jae had pushed it with the dinner invite. The thought of Marc in Jae's bedroom was clearly too much for Morel and he practically wedged himself between the two teenagers.

'You can go and listen to your records,' Morel said firmly. 'I think it's time for Marc to walk home.'

'Daddy . . .' Jae said pleadingly, but her father held his palm in front of her face, and there was a hint that this might become a slap if she protested further.

'Bed, now,' Morel said fiercely.

Jae looked close to tears as she swept out of the drawing room and thumped upstairs to her bedroom.

Morel pointed Marc out towards the hallway. His face was bright red as the grandfather clock became a countdown to doom.

'Is that my son's tie?' Morel growled.

'Jae lent it to me,' Marc said, as he freed the knot and handed it to Morel.

'You're a good worker, but that's *all* you are,' Morel said. 'You're a handsome boy and it's understandable that my daughter is fond of you. But she's fourteen. It's much too young to be carrying on like this. The next time she invites you somewhere, you will decline politely but firmly. Is that clear?'

Jae was the best thing in Marc's life and he felt a little surge of anger. But he wasn't brave enough to defy Morel inside his own house.

'Thank you for a nice dinner, sir,' Marc said meekly. 'I'm sorry I caused you trouble.'

CHAPTER TWENTY-THREE

Marc didn't tell the nuns that he wasn't working Friday. They firmly believed the old saying that *the devil finds work for idle hands* and he'd already spent several precious days off painting fences, clearing gutters and mowing lawns.

The lake was in a valley at the farthest end of Morel's cornfields, with a large hay barn and a thick hedge keeping it nicely out of view. Jae was waiting at the water's edge, wearing canvas shorts and a tight blouse.

She had a ground sheet laid out and a wicker picnic basket. The weather was about perfect: hot, but enough of a breeze to take the sting out of it.

'Hey there,' Marc said softly, as he sat on the mat beside her. It was the first time he'd seen her since dinner two days earlier, but Jae backed off when he moved in for a kiss.

'What's the matter?'

Jae mocked Marc's voice. 'Thank you for a nice dinner, sir. I'm *sorry* I caused you trouble.'

'Well, what was I supposed to say?'

'You could have shown my dad that you had backbone,' Jae said. 'You could have said you liked me and would never do anything to hurt me.'

Marc tutted. He'd been looking forward to a day off with Jae and hadn't expected to walk into an argument.

'Your dad's my boss,' Marc said, as he started unbuttoning his shirt. 'If he sacks me I could be sent away to work in the factories. Now I'm going in for a swim.'

Jae looked on impassively as Marc threw off his boots, dropped his trousers and plunged into the water. His whole body went into spasm and he yelped with shock before running back out.

'Christ, its freezing!'

Jae was still officially sulking, but couldn't quite hide her smile. As Marc jumped up and down while drying his shoulders on a ragged towel, Jae slipped off her canvas pumps.

'I'll race you,' she said teasingly. 'Touch the tree overhanging the far side of the lake and swim back again. If I win, you have to face my dad and tell him that you like me.'

Marc hadn't swum in a while, but he'd done a lot of training in the lake on CHERUB campus and fancied his chances.

'You're on,' Marc said. 'But if I win, you have to stop calling me a coward.'

Marc's jaw dropped as Jae took her blouse off, exposing her bare breasts. His face went bright red and he didn't know where to look.

'You like?' Jae said, thrusting her chest out as she dropped her shorts and knickers too.

'You're crazy,' Marc said, not knowing where to look.

'It's what skinny dipping's all about, isn't it?

Marc was in a real state: he'd been brought up by nuns, who'd taught him that sex, nudity and lustful thoughts were sinful. But Jae naked was about the most exciting thing he'd ever seen.

'Takes two to tango,' Jae said, as she lunged forward and yanked Marc's soggy undershorts down to his knees. 'Nice bum. Ready, set, *go!*'

Marc still had his shorts twisted around his knees as Jae waded into the water and started swimming.

'I'm not ready,' Marc shouted, as he kicked off his shorts and dived in. 'No fair!'

Marc was taller and stronger, but Jae had a beautiful swimming stroke and he couldn't catch up. The lake was less than twenty metres across and the last stretch to touch the tree involved wading a few paces through a muddy reed bed.

Marc doubted he'd catch Jae in the water, so he tackled her as she came out of the reeds. Drips of water poured out of Jae's hair and over her naked body as she stood up and squealed.

'Cheater!' Jae yelled, as Marc lifted Jae high out of the water and dropped her down, making a huge splash.

As Jae surfaced, she grabbed a handful of the hairs between Marc's legs and gave them an almighty yank.

'OWWWWWWW!'

As Marc doubled over, half giggling, half in agony, Jae pushed off and started swimming for home.

'So long, loser,' she shouted.

Marc had more stamina and nearly got hold of Jae's ankle before she got back to shore. They were both muddy and scratched from the reed bank as they collapsed breathless and naked on the ground sheet.

After pouting at one another for a few seconds, they embraced and started snogging like mad. Five minutes passed before Jae rolled away from Marc and pushed his hand off her breast.

'I wonder what Father Denis will make of this when I go to confession on Sunday,' Marc said.

'Don't you *dare*,' Jae said. 'Father Denis eats dinner at our house.'

'Confession is secret,' Marc said.

'But I'd have to look him in the eye,' Jae giggled. 'And besides, the randy old goat would probably enjoy thinking about it.'

'I wish we were older,' Marc said. 'We could just find a church and get married. Or run off into the woods and live in a cave. Somewhere the war can't reach us.'

'And what exactly are you planning to say to my dad?' Jae asked.

'Eh?' Marc asked.

'The bet you just lost,' Jae said. 'You've got to stop chickening out and tell my dad that you like me.'

'I always keep my promises,' Marc said. 'It's a question of finding the right moment.'

'I could invite you to Sunday dinner,' Jae said.

'That's a bit soon,' Marc said. 'I was thinking the seventeenth of July 1961. After all, I promised I'd speak to your dad, but I didn't say when, did I?'

*

Marc and Jae didn't say much for the next couple of hours. They lay side by side letting the sun dry their bodies. Then they got dressed and ate a picnic together. The village school turned out at 12:30 and on a warm day like today, the tranquil lake would be dive-bombed by screaming, splashing kids.

Saying goodbye to Jae made Marc's heart burst. Back at the orphanage, Sister Mary Magdalene wasn't happy that he hadn't told her it was his day off. Fortunately the old nun didn't know what he'd been up to, but she still gave him a three-and-a-half-hour-standing-in-the-hot-sun-scrubbing-thirty-sets-of-bed-linen-before-wringing-them-all-through-a-mangle-and-hanging-them-out-to-dry punishment.

As Catholics didn't eat meat on Friday, the nuns served fish soup, after which Marc decided that the best way to take his mind off Jae was accepting an invitation to go out hunting with the two eleven-year-olds, Victor and Jacques.

The Germans had made all firearms illegal, so the boys were restricted to setting traps and snares for small game and foraging for mushrooms, snails and berries.

Victor and Jacques were eager to hear Marc's fictionalised tales of runaway life in Paris, but he felt surplus to requirement as the younger boys led the hunt. They bagged rabbits, squirrels and even an escaped chicken from their expertly set traps.

Marc was no hunter and felt stupid when Victor caught him picking poisonous mushrooms.

'I'd stick to snails if you want to live,' Jacques teased.

But Marc enjoyed rambling through the countryside, getting muddy clothes and bloody fingers, while listening to the crude jokes of his younger companions. He was only three years older, but after all he'd been through he had little in common with two lads who'd never strayed more than ten kilometres from the orphanage.

The evening's prize catch was a big freshwater trout, still trying to swim its way out of an underwater snare.

'Marc may be a bloody useless hunter but he's definitely a good luck charm,' Jacques said, as he scooped the flapping fish into a piece of netting and raised it out of the water. 'We've caught a couple of fish in water traps before, but something always seems to bite lumps out of 'em before we get here.'

Marc knew Victor and Jacques were messing, but his pride was dented and while Victor gutted the fish and used some of its intestines to re-bait the water snare, he pulled the cook's knife he'd stolen from Großmarkthalle off his belt and pointed it up at a tree.

'Who needs traps?' Marc said. 'Squirrel, third branch up. Watch and learn.'

Marc sent the knife spinning through the air. The squirrel felt it coming and dodged before impact, but the blade still speared its back, nailing it to the tree trunk.

The squirrel shrieked desperately as its blood rained on the leaves below.

'Impressive,' Jacques said. 'But how *exactly* do you plan on getting your knife back?'

'Oh,' Marc said weakly, as Victor and Jacques shook with laughter.

CHAPTER TWENTY-FOUR

Marc thought about climbing the tree, but it looked flimsy. Shaking the branches did no good, but after a couple of minutes Victor threw a big rock, which knocked the knife out of the trunk and put the squirrel out of its misery.

Jacques was laughing so hard he could hardly stand up straight as he picked the mangy blood-soaked squirrel out of the undergrowth tail-first and waggled it under Marc's nose.

'I *really* feel such a grand kill should be stuffed and mounted on the wall above your bed.'

'All right, smart arse,' Marc said, as he wiped his bloody knife on a handful of leaves.

'I'll admit you're a good shot with the knife though,' Victor said, as they started walking again. 'I reckon this is our best haul ever. The nuns are gonna love us.'

'So it's more than a hobby?' Marc asked.

'We're the best hunters in the orphanage,' Jacques told Marc proudly. 'Sister Mary Magdalene said our traps made a big difference when food was short last winter.'

It was getting dark and after checking a few final traps, the three boys started walking along a pathway at the edge of Morel's wheat fields.

They weren't far from the lake where Marc had begun his day when Victor suddenly stopped walking and made a shushing noise.

'I saw something moving up there,' Victor said, pointing at a barn.

'Are you sure?'

The sound of a wooden door creaking confirmed any doubts.

'Would anyone be out here this late?' Jacques asked.

Although Marc had mainly been tagging along with the hunting, he was oldest and felt a sense of responsibility.

'It could be thieves,' Marc said. 'Or maybe just someone working late. Do you two know how to get to Morel's house from here?'

'Of course,' Jacques said.

'If you two head up there. I'll creep up to the barn and try working out what's going on.'

'They might have guns,' Victor said.

Marc nodded. 'I'll be careful. And don't you two make any noise when you're running.'

Jacques pulled a slingshot and a few pebbles out of his trouser pocket. 'You want this?'

'Can't do any harm,' Marc said.

'Be careful, Marc,' Jacques said, as he led Victor off towards the house.

Marc waited for the boys to clear out before setting off cautiously towards the barn, roughly forty metres away.

The terrain was mostly long grass, but Marc had to leap a drainage ditch. He slipped on the embankment, muddying the knee of his trousers, then kept low as he approached the barn. If it was thieves, Marc reckoned they would have grabbed a few tools and left already, and anyone working in the barn would have left the doors open because of the heat.

When Marc reached the side of the barn he could peer through gaps between the wooden slats. It wasn't easy, but he eyed a large man resting in the hay. He seemed to be wearing a dark grey soldier's uniform. There were two kitbags on the ground and the man appeared to be writing something in a notebook.

Marc remembered his encounter with Tomas a couple of days earlier, and what Morel had said about the Requisition Authority sending inspectors out to spy on his farm. Rather than confront the man, Marc decided to follow Jacques and Victor up to the house, where he could let Morel know what was going on.

But as Marc stepped back a tall figure sprang out of the long grass and bundled him against the side of the barn.

Marc kicked backwards, hitting the man in the gut. The man charged while he was still doubled over, butting Marc in the stomach and knocking him against the barn again. As he stumbled sideways, the assailant straddled him and ripped out a knife, holding it to Marc's throat.

He spoke French, but with a weird accent. 'Don't move a muscle.'

The man who'd been inside the barn had heard the noise and run outside. He pinned Marc's arms, while the guy with the knife grabbed his legs. They took Marc inside, threw him down in the hay and shone a torch in his eyes as he rolled over.

'Just a kid,' a man clutching his stomach said, 'But he's got a kick on him.'

'We're not going to hurt you,' the one with the torch said.

Marc was a little dazed. The men wore baggy linen shirts like a French farmer, but their trousers were commando grey and they wore rubber-soled American-style boots. The puzzle pieces formed a picture in Marc's head when he remembered some men he'd met in Devon almost a year earlier.

'You're Canadians,' Marc said. 'Did you land at Dieppe?'

'You heard about that?' the one with the torch said.

'I heard,' Marc said. 'And Hitler's order: any survivors from the Dieppe raid who don't surrender will be shot as spies.'

'We're not here to hurt you,' the one Marc kicked in the guts said. 'My name's Noah. We're trying to move south. But we need food, and maps.'

'I thought you were from the Requisition Authority,' Marc said. 'I sent my two mates up to fetch the farmer. There'll be coming any minute, so start packing your things up.'

'If it's just a farmer . . .'

Marc didn't let him finish. 'There's Germans billeted to live in the house too. There's a ditch back there, about twenty metres. It's muddy, but you get down in there and I'll cover for you.'

The two Canadian's didn't seem too sure about trusting Marc, but Noah peeked out of the barn door.

'Think I hear something, Joseph.'

Marc sat up and spoke with authority. 'If you get in the ditch *now* I can cover for you. Pick up all your stuff and *move*.'

Marc had no idea if the two Canadians would do what he said, but they started running towards the ditch and Marc strode towards the group jogging through the wheat field from the farm house. One of the Germans billeted at Morel's house led the way, brandishing a rifle.

Morel, Jae, Victor and Jacques were close behind.

'Don't shoot,' Marc shouted. 'It's me.'

It was half a minute before Morel was on the scene and had caught his breath.

'It was kids,' Marc said. 'Less than my age. I think they were playing, but they both scarpered when they saw me.'

Jae smiled at Marc as Morel opened the door of his barn and looked inside.

'Nothing even worth stealing in here,' he said. 'Looks like they busted my lock though.'

'Which way did they run?' the German asked in broken French.

Marc pointed away from the ditch. 'Down towards the lake. I'd have chased them, but I turned my ankle jumping over the ditch.'

'I'd better report the sighting,' the German said. 'They're still on the look out after the Dieppe raid.'

'Not much point if it was kids,' Morel pointed out. 'I'd rather not have my farm torn apart in a search. And no doubt they'll have us up half the night answering questions.'

'It definitely was,' Marc said. 'They were littler than me.'

'Well, if you're sure,' the German said. 'And seeing as I've got an early start tomorrow.'

Morel and the German turned back towards the house, leaving Marc standing beside Jae.

'It was really nice earlier,' she whispered.

Marc gave Jae a quick kiss on the lips, 'I can't stop thinking about you.'

Marc had whispered, but it was still loud enough for Victor and Jacques to hear and they both started to giggle.

'You looooove her,' Jacques howled, as Victor laughed so hard that he had to clutch his sides.

'I'd better go before my dad flies off the handle,' Jae said. 'I'll find you at lunchtime tomorrow.'

As Jae caught up with her father and the German, Jacques and Victor started picking up the sacks filled with dead rabbits and squirrels. Marc rushed to the ditch. He thought the Canadians might have run off, but they'd waited as he'd asked.

'Saved our bacon,' Joseph said.

Marc spoke quickly, in a whisper. 'I'm not sure I can trust the two younger boys to keep their mouths shut. But we're going cross country. There's no checkpoints. So follow us. When we go inside our orphanage, you two wait behind the wall at the back and I'll come out and meet you. I can definitely get you food and directions, maybe some better clothes. OK?'

Two sets of grinning teeth looked out of the muddy ditch. 'Thank you, son,' Noah said. 'I reckon we'll need all the help we can getting out of this scrape alive.'

CHAPTER TWENTY-FIVE

'Don't you like getting your hands bloody?' Victor asked Marc, as they stood alongside the burned-out barn at the side of the orphanage gutting a squirrel. Jacques was doing the chicken and Sister Raphael was helping out too.

'I twisted my ankle in that ditch,' Marc said, as he headed towards the orphanage. 'I need to sit down.'

He'd looked back a couple of times and seen the Canadians following. If they'd done what he'd asked, they'd now be crouching by the perimeter wall.

Marc stepped through the back door of the orphanage and headed for the kitchen. The nuns pulled every trick in the book to keep their orphans well fed, so stealing from the kitchen was a grave offence. Luckily at this time of day the nuns had their hands full putting the little kids to bed, so Marc was able to sneak some bread, a couple of eggs and a big hunk of cheese.

He had to go out of the front door and right around the outer edge of the orphanage grounds to avoid being seen by Jacques and Victor, but walked straight into Sister Madeline. She looked furious when she saw the food, but spoke in a whisper because she was cradling a toddler.

'Don't you wake him,' Madeline said sharply. 'He's got an ear infection and his screaming was keeping the others awake. What are you doing?'

If it had been any other nun Marc would have said he was scoffing the food for himself and taken a caning as punishment. But Madeline was a caring person. She'd always had a soft spot for Marc and had been the only nun brave enough to regularly stick up for him during Director Tomas's reign.

'I found two Canadians,' Marc said. 'They're dirty and hungry.'

'From the Dieppe raid?'

'I think so,' Marc said.

Madeline looked anxiously at the little boy in her arms. 'I'll put him to bed. You take the men to the oratory.'

'You can't tell anyone,' Marc said. 'There are notices up in Beauvais. Any men who didn't surrender immediately after the raid will be shot as spies.'

'We've dealt with this before,' Madeline said firmly,

and to Marc's complete surprise. 'Just do as I say.'

Marc realised he was in an awkward position as he ran around the crumbling limestone wall surrounding the orphanage. He was a trained espionage agent, but he was only fourteen and the nuns would expect to take control of the situation.

'Here,' Marc said, as he handed the two big Canadians some food. 'We're going to the oratory, next to the convent house where the nuns live.'

'Why are we going there?' Joseph asked, as he greedily scoffed bread and cheese.

'I'm not forcing you to do anything,' Marc said, as he remembered a line he'd heard during his parachute training the previous year. 'But if you're going to get home, you're going to have to trust someone. And if you're going to trust someone, it might as well be a nun.'

Joseph and Noah exchanged glances and muttered in English.

'He's just a kid.'

'He had the smarts to save our bacon at the barn.'

As far as Marc could tell, Joseph was a native English speaker, while Noah was native French. Marc put on his best attempt at a posh British accent and spoke in English for the first time in ages.

'I can still understand what you're saying, chaps.'

The two Canadian's found Marc's accent hilarious and the joke broke the ice.

'What was going on back there at Dieppe?' Marc asked, as he led the two soldiers across a couple of hundred metres of open country towards the small convent house.

'Whatever the plan was, it didn't bloody work[9],' Noah said, smiling uneasily as he scratched three days' stubble. 'Heavy machine guns covered the beach from every angle. We landed within the first hour, but you couldn't put your boot down without squelching a dead body. We spent three hours pinned at the base of a cliff, ran out of ammo and surrendered. We got captured, but took our chances and scarpered while the Boche were trying to work out where the hell to put all the prisoners.'

'Tough break,' Marc said.

They'd now crossed behind the orphanage and reached the small convent house, which was home to the six nuns who worked in the orphanage and two frail old

---

[9] The Dieppe raid – officially named Operation Jubilee – took place on 19 August 1942. It involved 6,000 Canadian and British troops. The aim was for a large force to capture the harbour at Dieppe, destroy German facilities, take key pieces of German technology and abduct senior officers for intelligence purposes, before withdrawing in an orderly manner.

The raid resulted in the loss of 96 Allied aircraft and 34 ships. 3,623 of the 6,086 Allied soldiers who landed on French soil were either captured or killed. It is regarded as one of the most disastrous Allied operations of the entire Second World War, although lessons learned in the failure at Dieppe were crucial to the success of the much larger D-Day landings which took place two years later.

sisters who rarely ventured out. Next door was the small brick-built oratory where each nun retreated for several hours of daily prayer.

Although he'd lived his first twelve years in the orphanage, Marc had never been inside the oratory, partly because the nun's quarters were off limits, but mainly because it was surrounded by a graveyard for orphans which always freaked his younger self out.

The octagonal prayer space was elegantly simple: whitewashed walls, with rough pine benches and a fireplace with a wooden cross resting on it.

The teenage Sister Peter knelt in flickering candlelight, studying a Bible as Marc and the two burly servicemen stepped in. The smell of three days without a bath quickly overpowered the incense, and the light gave Marc his first proper chance to study the two soldiers.

Joseph was stocky, touching thirty, with bushy, red hair and slightly sad eyes. Noah was a bigger man, no more than twenty but with huge hands and thighs as broad as Marc's waist.

'God be with you,' Sister Peter said awkwardly. 'You're welcome here, strangers.'

Sister Madeline was only a few moments behind and the two young nuns began fussing over the Canadians, bringing wine and more food, along with bowls of hot water so that the men could wash and shave.

Given that they'd travelled ninety kilometres from

Dieppe in under two days, the nuns gave most attention to the mens' injured feet, popping blisters and pushing straw into their soggy boots to dry them out.

While the nuns fussed over the soldier's immediate physical needs, Marc hoped to help them in other ways. The silenced pistols and detonators he'd noticed on their belts indicated that they were part of a commando team rather than just regular soldiers.

'You've done well to get this far,' Marc said. 'Do you have a plan?'

'Got lucky a couple of times. Near misses, with sniffer dogs and checkpoints,' Noah said, sounding every bit like a man who'd not slept in three days. 'First priority was to get away from Dieppe as fast as we could. Walking mostly, but we did jump on the back of a truck for a while.'

Marc nodded. 'You'll need civilian clothes to have any realistic chance. You've been able to sneak through countryside so far, but it starts getting built up not far south of Beauvais.'

Joseph looked wary. 'In uniform we can be taken as prisoners of war,' he said. 'We'll be protected under the Geneva Convention. Once we switch into civilian clothes they can shoot us as spies.'

'They might shoot you anyway,' Marc said. 'The Germans are pissed off about the raid, and you're a long way from Dieppe already.'

'Can you find us civilian clothes?' Joseph asked.

'We can all sew,' Sister Madeline said brightly. 'It's identity documents that are the problem.'

'You got mine easily enough when I came back,' Marc pointed out.

'Because we're an orphanage,' Sister Madeline said. 'Children arrive here regularly and we have special arrangements at the identity office. Requesting documentation for two adults is more suspicious.'

'It's almost impossible to move around anywhere close to Paris,' Marc explained to the Canadians. 'You get stopped at checkpoints two or three times a day.'

'A man gave us an address,' Noah said, only for Joseph to give him a look that suggested he shouldn't have spoken.

'What address?' Marc asked.

'We jumped off the back of a German truck, and ended up near Amiens,' Noah explained. 'We knocked on doors, but everyone was terrified of helping us.'

'The Gestapo send out spies pretending to be airmen, or allied soldiers,' Marc explained. 'If you help them, you're for the chop.'

Noah raised an eyebrow. 'So how are you so sure we're not Gestapo spies?'

'I knew a Canadian once,' Marc said. 'Plenty of Gestapo officers speak French, but not with those kooky accents.'

Joseph laughed. 'Kooky, eh?'

'Anyway,' Noah said, continuing his story. 'We finally found one old guy. He let us fill our canteens, gave us a little food, dry shirts and socks, plus the address for someone in Paris who he said might be able to help us.'

Marc took a crumpled piece of paper from Noah and unfurled it. He was horrified that the name *Chalice Poyer* and an address of an apartment in the 18th Arrondisment of Paris had been written down uncoded. If the two Canadians had been captured, the address would have been raided and its occupants tortured.

'Did the man say who Chalice is?'

'He was a farmer. Keen to help, but scared out of his wits,' Noah said. 'He just said that the girl was connected to people who might be able to help us.'

'He didn't say how he knew her?' Marc asked.

'The old man had copies of this anti-Nazi newspaper,' Joseph said. 'Not even a newspaper really, just a typed sheet. I think he wanted to help us more, but he had a daughter and three young grandchildren in the house. His hands were trembling.'

'I've seen anti-Nazi newspapers,' Marc said thoughtfully. 'People leave piles of them in Metro carriages. Perhaps this Chalice is involved in distributing them.'

Marc had made no effort to unearth any local resistance activity since he'd arrived back at the orphanage. Partly it was because he doubted there was any around, but mainly because he'd found a degree of comfort, with a tolerable job, plenty of food and his burgeoning relationship with Jae.

'I met an RAF fellow in Britain who said there were lines of resisters that regularly help downed aircrew escape into Spain,' Joseph said. 'Do you think this girl could have links to them?'

'That address could be anything from the home of a genuine resistance leader, to a false address circulated as a Gestapo trap,' Marc said warily.

Noah looked at Sister Madeline, who was rinsing out his socks in the brown water he'd used to wash his feet. 'What do you think, sister?'

'We are brides of Christ, not worldly creatures,' Madeline said apologetically. 'We care for orphan boys and whoever else the lord sends our way. An airman passed through in similar circumstances to yours last winter. We fed and sheltered him for a short period, but we have no idea what happened after he left us.'

Marc saw Sister Madeline's words as an opportunity to assert control over the situation, without having to tell a scarcely believable story about being a trained espionage agent.

'I know my way around Paris,' Marc said. 'You two have no documentation. You speak French well, but your accents are unusual and men of fighting age are always treated with the highest suspicion. I'd suggest that we find a place to hide you. You can sleep and rest. The nuns can find you civilian clothes and food. This girl is our only lead and if Sister Madeline allows me, I can travel into the 18th Arrondisment and check out this address.'

'If it's a trap, they'll arrest you,' Noah said. 'And you'll lead them straight back to us.'

Marc nodded. 'I'll have to be careful. But I've spent the last two years living off my wits in Paris. I won't just knock on the front door and ask this girl for help.'

'It's still risky,' Joseph said.

'Everything's risky,' Marc said. 'I'm willing to try helping you. Frankly, I'm amazed you got this far dressed in uniform and with no documents.'

'What if the girl's no help?' Joseph asked.

Marc shrugged. 'We'll have to think of another way of getting documents and helping you move south towards Spain.'

Sister Peter looked sympathetically at Marc, and spoke in a gentle voice. 'He left here and survived for two years, when no other runaway lasted more than a few days. I sense the hand of God in you three coming together like this.'

The two Canadians didn't look keen on entrusting their fate to a fourteen-year-old, but they were exhausted and didn't have many options.

'Get down to Paris then, I guess,' Noah said, smiling

at Marc. 'I'm not gonna be walking far these next few days anyway.'

'I'll sort you food and money,' Sister Madeline told Marc. 'I'll send one of the younger boys to tell Felix that you're unable to work tomorrow.'

'I may have to stay overnight,' Marc said. 'If the girl works in the day, for instance. I may not be able to speak with her and get back here before curfew.'

'I'll bring blankets and pillows so that Noah and Joseph can sleep here in the oratory tonight,' Sister Madeline said. 'We'll find somewhere more secure before daybreak. But right now, I think we should put our hands together in prayer and seek the lord's guidance.'

Joseph nodded, and gave Marc a slight smile. 'Can't see no harm in having God on our side.'

CHAPTER TWENTY-SIX

Marc couldn't sleep. Swimming with Jae and cuddling up wet and naked afterwards was one of the most beautiful, mind-blowing things that had ever happened to him.

He'd spent most of his life stuck in this orphanage longing for adventure. Now he wanted nothing more than to stay where he was, working on Morel's farm and spending time with Jae, but adventure had been thrust upon him.

Marc didn't feel like getting up early and going to Paris. His escape had been a dangerous game, which he'd won as much by luck as skill: landing the job in the Labour Administration office, Fischer being a terrible shot, the friendly gendarme, the girl who could have screamed when he stole the German's wallet.

Gambling with your life doesn't matter when you've got nothing to live for, but now he had the memory of Jae's breasts pressed against him, the baby-fine hairs on the back of her neck and her delicate fingers. He'd had crushes on girls before, but he was certain this was love.

Yet Marc felt for the two Canadians. In Germany, he'd tended to think day-to-day: how to avoid a nasty guard, how to keep bugs off his mattress, what would the next meal be? But when he thought about the prisoners now, he kept wondering how it would end for people like ex-cabin mates Vincent and Richard. If the Allies invaded Europe and put the squeeze on Germany, the prisoners would get worked harder and fed less until they died. If the Germans won, they'd live out their lives as slaves.

So when Marc jumped off his bunk at 5 a.m., he resented having to re-enter the world of dodgy documents, chases and checkpoints, but he genuinely wanted to re-establish his links with the resistance and help the two Canadians.

'Why are you getting dressed?' Jacques asked, from the bunk below.

'Go back to sleep,' Marc whispered.

Down in the kitchen, Sister Madeline had laid out bread and cheese for Marc's breakfast, a canvas bag with a sandwich and some apples for lunch, plus ration coupons and money in case he needed to stay overnight.

Marc added a change of clothes, a small metal file and purposely-bent piece of wire that would serve as a basic lock-picking kit, plus the cook's knife he'd used to kill the squirrel the night before.

The sun was starting to come up when Marc finished a brisk six-kilometre walk from the orphanage to Beauvais Station. The seventy-minute train ride was quiet, but hordes of Parisians crowded the platform when his train reached the city centre, waiting to board the train going the other way.

Spending a weekend in the country north of the city had always been popular with Parisians, but tight rationing added the illicit attraction of buying black-market food from farmers.

Chalice Poyer's address should have been a short Metro ride, but the line was closed so Marc had to take two buses. He stepped off in a hilly neighbourhood of five- and six-storey apartment blocks: the kind of place where an office secretary or minor civil servant might live.

The buildings had names rather than numbers, and Marc had to get directions at a cafe. Chalice Poyer lived in apartment 3–4, on the fourth floor of a narrow block. Like most Paris apartment blocks, there was a desk for a concierge, but judging by the dust and the absence of a chair nobody had worked this desk in years.

Marc checked the row of metal boxes in the lobby. The

name *Poyer* was still on the mailbox for apartment 3–4. Although the box was locked, there was enough of a gap for him to see a couple of letters inside, but not enough to make him think she'd moved away.

If Chalice had lived in a house, Marc could have skirted around, peeked through the windows or looked over the back wall. Checking out a fourth-floor apartment was trickier.

He passed a woman on the stairs who took no notice of him. Marc had thought up a cover story while he was on the train. If anyone asked his business he'd say that he'd lived nearby when he was younger and was searching for an old school friend who he'd not seen since the invasion.

The fourth floor had five apartments. Number three was at the end of a short corridor, with a set of metal fire stairs crossing the window at the far end. Little kids played rowdily in the apartment opposite, and a whiff of drains and urine came from the communal toilets and rusted bath tub shared by all five flats.

Charles Henderson had taught Marc that the critical task when making an approach is to be patient, and find out everything possible about your target before letting them know you exist.

Marc's first task was finding out whether Chalice was at home. He knocked loudly on the door, then backed quickly into the communal bathroom across the hallway.

The door didn't open, so she was either out or a heavy sleeper.

Marc decided that the fire escape would be his next step. Opening the window made more noise than he would have liked, but the metal balcony brought him within reach of a narrow apartment window, which had mercifully been left open in the hot weather.

To be certain nobody was inside, Marc gave the window a noisy upwards shove, then backed up to the wall. When no head popped out to investigate, Marc stepped over the balcony on to a precariously fragile window ledge and crawled through, dropping down on to polished wooden boards.

The apartment was a single room, with a built-in wardrobe and wash basin along one wall. The only other items were a metal-framed double bed and a chest of drawers. Everything was neat, which was how resistance operatives were taught to be, because it's easier to detect if a neat apartment has been searched than a messy one.

Marc was wary of touching anything, but he noted a distinct lack of personal mementos, such as photographs or address books. This was another sign of a trained spy, because if you didn't crack under torture, the Gestapo would try and break you by threatening friends and family.

Now that Marc's initial wariness had worn off, he peeled back and carefully replaced the bed clothes. The sheets were cold, as was the kettle on the stove and the dregs in the coffee cup on the bedside table.

Next he looked in a couple of drawers and opened the wardrobe. It seemed Chalice had a slim figure, along with a taste for good clothes and makeup. There was some expensive-looking evening wear, but also sets of plain brown dungarees with ink stains and a couple of wage slips for a job with a printing company.

The printing company tied up with the Canadian's theory that Chalice was linked to the anti-Nazi newspapers he'd seen in the house in Amiens. The date of birth on a wage slip indicated that Chalice was only nineteen, but she wasn't earning enough for the nice clothes so Marc reckoned there had to be a rich daddy or rich boyfriend in the mix somewhere.

Marc had clearly unearthed an interesting character, but the door rattled before he could investigate further. A turning key meant it was almost certainly Chalice, but Marc was saved from a confrontation by the double locks.

By the time Chalice was inside, Marc had scrambled out on to the fire escape and the only clue to his entry was that he didn't get a chance to push the squeaky window back down.

Chalice was attractive, with dark tangled hair and a tight-fitting black dress. The look suggested she'd spent Friday night on the town and slept elsewhere. She'd brought in a fresh, black loaf, a newspaper and the letters from the mailbox downstairs and sat on the edge of her bed dunking the hard bread into a glass of milk.

Marc thought about knocking on the door and introducing himself, but heard Charles Henderson's voice in the back of his mind, telling him to be patient and act on solid information rather than gut feeling.

Half a day following Chalice would make no difference to the Canadians, but sooner or later someone would spot Marc skulking on the fire escape, so he retreated into the communal bathroom and bolted himself in a cubicle.

An old boy stunk the place up taking a huge half-hour dump and the woman from the apartment with the little kids came in and started yelling at him.

'What am I supposed to do, shove a cork up my arse?' the old man roared.

'You should see a doctor,' the woman shouted back, as her toddlers stood behind her legs pinching their noses. 'That smell isn't natural. It seeps through to my kitchen.'

Marc would have seen the funny side if the stench hadn't been filling his nostrils. And the shouting match almost made him miss Chalice leaving her room.

She'd tied her hair back and changed into ink-stained printers' overalls. Marc crept down the fire stairs and emerged on the doorstep seconds before Chalice vanished around a street corner. He kept his distance as he followed her down a couple of narrow streets to a bookshop.

Chalice was only inside for a few seconds, but emerged from the shop with a different package to the one she'd walked in with. Then she cut through a small garden square and glanced around casually before pushing a ground spike into the earth between some shrubs.

Marc had never used a ground spike, but he'd heard about them in training. They were simply a hollow metal spike into which you placed a rolled-up message. Then you pushed it into soft ground, either earth or longish grass, and it would remain invisible until someone came along to dig it out.

Marc was now completely convinced that he'd found someone linked to an anti-German resistance organisation, but also a little concerned. The security in Chalice's apartment was impeccable, but Marc had followed her for half an hour and not once had she doubled back, or taken an indirect route to see if she was being followed.

Still, Marc knew from his experience in Lorient that most resistance groups were sloppy, and even well-trained agents grew complacent and fell into routine patterns.

Chalice's next stop was a tobacco stand, where Marc saw nothing beyond an innocent purchase of cigarettes. Fifty metres more took her through an arched, blue doorway into the premises of the Constellation Print

Works. She gave friendly nods to women in identical overalls who stood smoking in the doorway.

Marc knew from Chalice's payslips that her shifts lasted ten hours, with thirty minutes docked for lunch. It was just before noon, so in theory she'd finish work at around ten that evening. She'd be tired and Marc would have the advantage of surprise when he approached her.

CHAPTER TWENTY-SEVEN

Ten hours is a lot of time to kill, but it gave Marc a chance to investigate Chalice in more depth.

When he got back to the little garden, someone had already picked up the ground spike. A dead drop is an excellent way for espionage agents to exchange information without meeting each other, but the less time you leave your message lying around the less chance there is of someone else accidentally finding it. The fact it had been picked up swiftly was another sign that Chalice was part of a reasonably sophisticated resistance operation.

She'd clearly swapped documents or information in the bookshop, so Marc decided to check this out next. The place sold a few new books, but it was mostly second-hand, with closely spaced shelves going from floor to ceiling.

Three women sat behind the counter at the back of the store, which was far more staff than the number of customers demanded. They gossiped about the things you'd expect: husbands and daughters, a touch of luck arriving at a butcher's shop at the same time as a delivery of lamb.

It didn't take Marc too long to work out that they were sisters – or possibly sisters-in-law. Evaline was the oldest by a good ten years. Halette and Julie looked alike, though Julie was plumper and had much shorter hair.

When a German came in, Evaline pointed him politely towards a small section of books in German, but when he left she slammed the cash draw and spat on the floor.

'Dirty Boche money.'

For the past two years, Marc had lived adventures rather than read them. But before the invasion books had been one of the few escapes from the dull life of an orphan, and he voyaged into his past opening books and reading paragraphs from stories he'd enjoyed when he was younger.

'Can I help you to find anything in particular?' middle sister Halette asked.

She had shoulder-length hair and a homely face. If she'd been an actress, she would have spent her career playing nurses, or the woman who answers the door when the policeman comes to say that your husband got killed in a car crash.

'You've got lots of nice books,' Marc said. 'But I don't have much money.'

Marc spent a pleasant half-hour talking with Halette about books he'd read, and books that she thought a boy his age might enjoy. While they spoke, nobody bought anything, but several people swapped books at the counter as Chalice had done, while Marc pretended not to notice.

Each time a book arrived, one of the sisters took it out back and spent several minutes doing something with it. Marc's best guess was that the bookshop was being used by various agents to drop messages, which the women were then taking out back to encode, ready for radio transmission back to Britain.

Either one of the sisters was a radio operator, or if the resistance circuit was more security conscious, the women would leave the coded messages in some kind of dead drop to be transmitted to Britain by a radio operator who they'd possibly never even met.

Eventually, Halette told Marc the name of a good local cafe and sold him a copy of *The Three Musketeers* for a nominal price. She told him it was the tattiest of the four copies she had in stock and she was glad to be rid of it.

It turned out that the cafe was run by a cousin of the three sisters. Considering the food situation, Marc's ration coupons bought him a reasonable lunch of hot soup, and the cafe owner didn't object to him eating the sandwich he'd been given by Sister Madeline.

After enjoying the first few chapters of his book sitting outside the cafe, Marc thought it would be interesting to go back to the little park, and see if he could spot any further shenanigans with ground spikes.

It was extremely warm. A bunch of nine- to eleven-year-olds playing football gave Marc no trouble as he lay in the grass, reading bare chested with his shirt rolled as a pillow under his head. It had been ages since he'd been able to sit and read, but although he liked the book his mind kept drifting towards thoughts about Jae.

He knew she'd be working on the farm today. In this heat the sweat would be pouring down her neck. He recalled the earthy smell she had when she'd been working hard, and the way she had to keep pulling the straps of a man's farm overall up on to her skinny shoulders.

Besides missing Jae, Marc found ways to torture himself. He got paranoid at the unself-conscious way she'd stripped off in front of him, and hugged him naked. What kind of girl did that? One who'd carried on with one of the older farm hands? Or with one of the German officers who lived in her house? Had Jae fallen for Marc like he had for her, or was he making a massive fool of himself?

'Why are you here?' a boy asked.

He was one of the footballers. Holes at the knee of his trousers and canvas pumps so tatty that several older siblings must have grown through them before they'd ever graced his feet.

'What's it to you?' Marc asked.

The boy shrugged and didn't sound unfriendly. 'We know all the kids round here. Did you just move in or something?'

'I came to meet an old friend,' Marc said. 'But he's working until later.'

'Right,' the kid said.

As the nosy kid went back to playing football, Marc spotted Chalice and two other women helping to load posters on to a cart outside the print works a couple of hundred metres away.

He was too far away to read the text, but he'd seen the poster all over Paris: The picture was of a smiling French gendarme, shaking hands with a smug-looking German soldier and the slogan *Working together to keep France safe*.

*

As Chalice wasn't scheduled to finish work until ten, Marc faced a problem with the curfew. The good news was that this quiet suburb didn't seem to be of much interest to security forces and he'd not seen a soldier or a gendarme since stepping off the bus.

But if things didn't work out with Chalice he'd need a place to stay overnight. There were no dormitory houses in the area, but there were quite a few empty apartments around and a waitress in a cafe directed him to one of the area's better buildings, where the concierge said he was welcome to sleep on the floor in a basement laundry room, provided he paid her two francs and behaved himself.

She seemed lonely and after finding Marc a couple of old sofa cushions to rest on, she babbled for ages about the patisserie she'd run with her husband, which had apparently been one of the best in Paris but had to shut down because her husband was sick and she thought it was a crime to cook with saccharin and powdered eggs.

'When this war's over, I'm gonna bring my girlfriend to your shop and we'll sit outside and eat the biggest, fanciest thing on the menu,' Marc said.

'Ahh, you have a girl, how sweet!' the woman said, laughing. 'No surprise really, you're a very handsome boy.'

But the woman was less pleased half an hour later when Marc said he had to pop out.

'But the curfew!' she said anxiously, before snapping her fingers. 'You be careful. They'll put a boy your age on the train to Germany like *that*.'

Marc gave a cocky smile and told the woman he'd be fine. Chalice's building was less than a hundred metres away. It wasn't yet ten, so he decided to wait in the communal toilet. He struggled through a couple more chapters of *The Three Musketeers* while listening out for the door of Chalice's apartment, but he was too tense to concentrate.

It was about quarter past ten when Marc heard footsteps in the hallway. It was more than one person, and as he leaned towards the locked cubicle door to try and hear more clearly, someone booted it from the other side.

The door smacked Marc in the head. He crashed back against mildewed wall tiles as a boot hit him in the guts, doubling him over. It was two big men, plain clothes. Both spoke fluent French.

'Open your mouth,' one ordered.

Marc tried to wriggle, but the man grabbed him around the neck and clamped a hand over his nostrils. As Marc gasped for air, a rubber ball got rammed in his mouth. Then he got a hard punch in the lower back, and saw the jagged edge of a butcher's knife at his throat.

'I will not hesitate to kill you,' the knifeman said. 'One wrong sound, one wrong move. You understand?'

CHAPTER TWENTY-EIGHT

Marc glimpsed Chalice in her spattered printer's overall as the two men pulled him out of the toilet cubicle and hauled him upstairs to the sixth floor. They took the knife away from his throat as they made him climb into a loft, then out on to a flat section of roof.

With a man in front, a man behind and Chalice at the rear, Marc was marched across the rooftops, struggling for breath with the rubber ball pinning his tongue to the roof of his mouth. After clambering over a set of chimney pots he was forced to jump down on to a metal fire escape.

They clanked rapidly down twelve flights of metal fire stairs. Marc looked back and tried to say, 'I can't breathe,' as he was forced to move at speed, but the man behind cuffed him across the back of the head.

After being picked up by the waistband of his trousers and thrown over a crumbling garden wall, Marc was hoisted out of a mix of mud and fire ash, then thrust through an open back door into a basement room with a heavy butcher's block table and a smell of damp.

Marc saw Halette from the bookshop standing at the bottom of a staircase as one of the big men bent Marc over the table. There wasn't much light, but Marc was close enough to the rough wooden surface to see that the tabletop was spattered with fresh blood.

'Who are you?' Chalice shouted.

Marc couldn't answer because of the gag. As one of the men smashed a large wooden club against the tabletop deafeningly close to Marc's ear, the other pulled his trousers and undershorts down to his ankles.

'Who are you?'

Marc still had no way to answer. The cook's knife he'd tucked into his belt clanked against the ground, and his bag was thrown across for Halette to inspect.

The bat smashed the table again, spattering some of the pooled blood over Marc's face as the other man finally ripped the rubber ball out of Marc's mouth.

He took three huge gasps of air, before the man with the bat pulled him up from the table. He brandished the bat under Marc's nose and screamed right in his face.

'One lie out of your mouth and I'm gonna stick this up your skinny pink arse. Got it?'

Marc was trembling, and still catching his breath as he nodded.

Across the room, Halette had switched on a desk lamp and stood carefully inspecting Marc's identity documents and the contents of his bag.

The man who didn't have the bat spat in Marc's face, then pointed at Chalice. 'Which German bastard sent you here?' he shouted. 'I want his name, *right* now.'

'There's no German,' Marc said, as a loud shudder entered his voice.

'Bullshit,' the man roared. 'You were seen. You snooped around her flat. Then you followed her. You went to the park. You went to the bookshop. Why are you spying on us?'

The one with the bat yanked Marc around and took over the shouting. 'Lowlife scum. I don't care if they're holding your mother and your baby sister hostage. You're still betraying your country. Tell me every detail and we might let you live.'

Chalice grunted. 'Or at least, you might not have to die too painfully.'

'Marc Kilgour,' Halette from the bookshop said. 'Is that your real name? These identity documents look very new.'

Marc jumped as the bat slammed the bloody table again.

'You wanna end up like the last poor bastard we

brought in here?' the other man asked, as he made his point by dabbing a huge nicotine-stained finger into the pooled blood.

Marc had been trained in how to deal with torture and thought he was pretty tough, but this was complete sensory overload. His hands and legs were shaking and tears streaked down his face.

'Please let me explain,' Marc begged.

The man with the bat placed the end under Marc's chin. 'This is my lie detector,' he said. 'Wanna guess what happens when she detects lies?'

'I came from Beauvais,' Marc stuttered. 'There are two Canadians at my orphanage. They got Chalice's name and address from a man in Amiens who said she may be able to help.'

Halette looked at Chalice. 'Who would know you in Amiens?'

'I have a great-uncle there,' Chalice said.

The man with the bat turned his spite on Chalice. 'How in the name of Christ does your uncle know what you're doing?'

'He knew early on,' Chalice said, as she eyed Marc suspiciously. 'Maybe the Gestapo found something out through him. Oh shit . . . I hope he's OK.'

'Why are you talking to your family?' the bat man shouted.

'I wasn't bloody talking to my family,' Chalice shouted

back. 'But people aren't stupid. Our families know what we do.'

Marc was relieved that some of the pressure had turned on Chalice, but that ended swiftly when the heavy not holding the bat slapped his face and pushed his cheek down into a particularly bloody portion of the table.

'So you come to town. Why not just knock on Chalice's door? Why follow her? Why go to the book store? Why risk breaking curfew?'

'I wanted to check her out,' Marc said. 'The Canadians said the Gestapo set traps.'

'Why did you come? Why not them Canadians?'

'They've got no documents.'

Halette stepped closer. She held the file and a piece of carefully bent wire under Marc's nose. 'What were you planning to use that for?'

'You could pick locks with it,' Marc admitted.

Everyone in the room gasped. The guy with the bat swung it hard against the back of Marc's legs. He rolled off the table and landed at the boots of the other thug. He was then grabbed off the floor, and slammed down on the table again.

'I told you my wooden friend doesn't like liars,' the bat man roared. 'Since when do orphan boys from Beauvais stake out suspects, watch dead drops and know how to pick locks?'

'Who knew the Gestapo trained 'em this young?' Halette said. 'But it makes sense. Who'd suspect his little blond head?'

Chalice started to sob. 'Oh Christ, I'm going to have to move. And my great-uncle? What if they've done something to the little kids?'

Marc realised the innocent orphan act wasn't going to wash. But the truth that he was an underage spy trained by the British didn't strike him as a tale that would make the bat-wielding manic happy either.

'I say we kill him now,' the batless thug said. 'We've had eyes on him since he first went into Chalice's back window. He's not been in touch with the Gestapo all day. All we can do is dispose of him.'

'Maybe we should check with Ghost first,' Halette suggested. 'If the Gestapo are training kids as spies, he may know things. Ghost may want him professionally interrogated.'

'Listen,' Marc gasped desperately. 'I have learned a bit about espionage. I know how to pick locks. I know how to follow people, speak German, send Morse, work a code book. But I didn't learn it from any Gestapo.'

Chalice wagged her inky pointing finger in Marc's face. 'Whatever they've done to my great-uncle, we'll do to you. Multiplied by ten.'

'Please,' Marc shouted. 'I'm *begging* you to listen to what I'm saying. You must have a radio operator in your circuit. Transmit my name. Tell them I worked with Espionage Research Unit B and got captured in Lorient last July. I don't know your transmission sked, but it shouldn't take more than a day to get your answer. If I'm lying, you can put a bullet through my head.'

'I'm not listening to this shit,' bat man roared. 'Kill him, cut him up. I'll put the bits through the sausage grinder.'

'No,' Halette said firmly. 'If the Gestapo are training kids to spy on us, he needs proper interrogation. This *has* to go through Ghost.'

'Bloody Ghost,' the bat man shouted, as he grabbed a handful of Marc's hair and banged his head on the table. 'What is it you want to know? I'll interrogate this pale-arsed collaborator right now.'

'You're not a trained interrogator,' Halette said. 'Chalice, you'll have to stay away from your apartment. Don't go to work tomorrow. We might have to get you new documents, but for now the important thing is to stay calm.'

'What about my uncle?' Chalice sobbed.

'Your uncle's fine,' Marc said. 'Your uncle's still in his house in Amiens. There *are* two Canadians from the Dieppe raid at my orphanage and if you contact London, they'll confirm my identity. I swear on my *life* I'm telling the truth.'

The thug put down the bat and grabbed a piece of

rope. 'How can you swear on a life that isn't worth shit?' he shouted. 'All that's left of it, you're going to spend in pain.'

CHAPTER TWENTY-NINE

The two thugs put the gag back in, and this time they tied it in place with a hair ribbon. The guy who'd held the bat put himself in charge of all the knots. He flipped Marc on to his chest and bound him tight, right wrist to left ankle, left wrist to right ankle.

Marc didn't want to give him the satisfaction of showing his pain, but he couldn't help moaning as the big men carried him up six flights and dumped him face first on the floor of a tiny loft room.

'I'm gonna be right below you,' the man warned. 'If I hear so much as a fly fart . . .'

Marc understood that his own side was suspicious. The fact that he'd been spotted as soon as he'd arrived showed that the resistance circuit he'd unearthed had decent security in place. But that didn't make it any nicer being tied up in a painful position and shoved in a dark loft amidst dead spiders and mouse turds.

The heat made it a constant struggle to breathe and as the night went by he had no option but to piss himself when he couldn't hold on any longer. This treatment seemed unnecessarily cruel and it depressed him to think that not all the bad guys were walking around wearing swastikas.

It was easy to fall into the trap of thinking that everything would be great if you kicked the Nazis out, but there'd still be plenty of cruel bastards left whichever side won the war.

*

There was enough light leaking between the rafters for Marc to know it was daybreak, but it was at least mid-morning when Halette from the bookshop climbed up into the loft.

'Don't bite me,' she said firmly, as she squatted amidst the dusty rafters and pulled out the ball. 'Show me your teeth.'

Marc scowled as the woman inspected his mouth. At first she didn't look happy, but then she tapped all his front teeth and disappeared back through the loft hatch without putting the gag back in.

'Don't start yelling,' she said firmly. 'Or I'll send the men up to sort you out again.'

'Can I get some water, at least?' Marc begged.

Halette didn't answer, but an hour later she came back with a knife and started untying the knots. Marc was bruised and filthy. He had awful cramp in his knees, and rope burns around wrists and ankles from being tied up.

'Are you going to tell me what's going on?' Marc demanded, then gasped with relief as he drained an enamel mug filled with cold water.

'According to our contact with Ghost, Marc Kilgour is a registered British intelligence source who had a tooth pulled by a Gestapo officer two summers back. That's why I was sent up to check your mouth.'

This really intrigued Marc. Unless things had changed in the year he'd been in prison, resistance circuits only transmitted coded radio signals once a day, so if you asked a question you'd always wait at least one full day to get a reply. The fact that Ghost knew about his missing tooth, meant that Ghost or someone close to him had met Marc before.

Getting down the loft ladder was excruciating, as his knees had been tied behind his back. After allowing him to wash and use the toilet, Halette led Marc down another floor, where he found himself in a dark butcher's shop.

The two men who'd terrified and beaten him the night before wore striped aprons and were dealing with a queue of women eager to buy either a revolting mound of sausage meat, or one of the rabbits hooked up behind.

As it was a Sunday and the shop's shutters were closed, it was clearly black-market trading.

Marc had been disorientated by the rooftops and alleyways the previous night, but the butcher's shop was only a few doors away from the bookshop.

'So what happens now?' Marc asked Halette, as they crossed the quiet street.

'You'll have some breakfast with my sisters in the back of the shop,' Halette explained. 'After that, Ghost wants to meet you personally.'

The bookseller made this sound like something extremely special, which only intrigued Marc further.

'So who is Ghost?'

'We don't know,' Halette explained. 'We use dead drops, false names and keep contact between resistance cells to an absolute minimum. The less we know, the less we can give away if the Germans catch us and torture us. Ghost leads the resistance in Paris. He's something of a legend. Some people say Ghost is not real, or that it's a leadership council rather than a single person.'

Marc was taken into the room behind the counter, where the three sisters had put together a late breakfast, with some almost edible bread, a large piece of cheese and some jars of home-made jam. In the country, most people could lay their hands on extra food, but Marc realised this was a good spread for Parisians and that it had been put on in an effort to apologise.

'We're sorry about last night,' Evaline, the oldest sister said. 'The butchers enjoy violence a *little* too much for our taste, but we've known them all our lives. We trust them absolutely and sometimes muscle is required at short notice.'

'Trust is everything in our line of work,' Marc said diplomatically. 'My sniffing around must have alarmed you.'

He wondered exactly what kinds of resistance activity the sisters were involved with as he spread a thick layer of butter over his bread. But he knew better than to ask, and if he had the sisters shouldn't have told him.

'I *really* hope you can forgive us,' Halette said.

But Marc had been badly knocked about and spent ten hours in pain. While he understood the reasons behind what he'd been through, he still hurt in too many places to feel like offering complete forgiveness.

'My trousers stink of piss,' he said trying to sound matter of fact, but unable to hide his bitterness. 'And explaining rope burns on my wrists will be interesting if I'm stopped at a Nazi checkpoint.'

The youngest sister Julie placed a neatly-tied bundle of books in the centre on the table. It was five of the volumes that Marc had discussed with Halette the afternoon before. They were all good-quality second-hand editions that would probably cost more than Farmer Morel paid Marc for a fifty-hour week.

'We don't have any men's trousers. But that's a small token of our apologies.'

*

Marc's instructions were to walk to Porte de Clignancourt Metro station, making sure that he took the first train leaving towards central Paris after 12:17 p.m. He was to travel in the last carriage but one and leave the train at Odéon, where someone would meet him and take him on to meet Ghost.

But the request to board a specific train and sit in a specific carriage made Marc suspect that he'd actually be met on the train. With Metro carriages unlit in the tunnels, Marc felt suspenseful every time the train rolled out of a station into darkness, but the voice that came at him as they left Gare Du Nord brought a lump to his throat.

'So it's really you,' Maxine Clere said quietly.

Marc gulped and pushed a tear from the corner of his eye. He was fond of Maxine and overwhelmed by the fact that he'd finally reconnected with his secret life.

Marc had first met Maxine when she worked at the British Consulate in Bordeaux, two years earlier. Maxine had devoted months to running an orphanage, helping reunite kids who'd been separated from their parents during the German invasion.

After returning to Britain, Maxine had trained as an espionage agent. Marc knew she'd been one of the first activists with the anti-Nazi resistance in Paris, but had no clue that she was Ghost, one of its most important leaders.

'I knew you were taken to Frankfurt,' Maxine explained. 'There was a possibility your espionage role had been unearthed and a double sent in your place, so I had to meet you in the flesh. We get off here.'

They left the Metro at Gare de l'Est, saying nothing more until they were above ground. Maxine led the way through the staff entrance of a restaurant directly opposite the station and up a narrow staircase to a small office. Marc felt stunned and slightly off-kilter, as if he expected to wake up and find he'd been dreaming.

'So you're the big boss now!' Marc said, as Maxine engulfed him in a hug.

'Until they catch me,' Maxine said.

There was a sadness in the way Maxine said this, and Marc noted that she had grey hairs, and shabbier clothes than the sexy creature who'd been seduced by Charles Henderson two summers earlier.

'You've grown up, you look healthy,' Maxine said.

'I looked like a stick when I got back from Germany,' Marc said. 'I've put on twelve kilos since the nuns started fattening me up.'

'My identity as Ghost is *absolutely* secret,' Maxine said. 'You must have no further contact with the women in the bookshop, or anyone else you saw yesterday.'

'Of course,' Marc said.

'We'll need full names, service numbers and all the information you have on the Canadians, which we'll transmit back to Britain. Once we've verified that they're not Gestapo spies, I'll set things in motion for return journeys.'

'Sounds good.'

'Full verification and document preparation takes time. Can the Canadians be looked after?'

'For a few weeks I expect,' Marc said. 'The nuns mentioned that they helped someone before. If you don't mind my asking, what route will you use?'

'The usual escape routes are over the Pyrénées into Spain; we have reliable escape lines and mountain guides taking out several dozen airmen every month. But our routes are currently overrun after the Dieppe cock-up, so things may take longer than usual.'

'What about *Madeline II*?' Marc asked.

'Sea crossings have become more difficult; the Germans are building up their costal defences. Lorient has become too dangerous, but *Madeline II* still makes occasional runs to the Brittany coast. And of course, I'll let Henderson and everyone else know you're safe and well.'

'Are any of them still in Lorient?'

'I don't know who's working in Lorient now,' Maxine said. 'And I don't ask, obviously. But I know that

Henderson returned to Britain. Safe and well, as far as I'm aware.'

'Feels good to be back in touch with everyone,' Marc said.

'I'm glad you're happy,' Maxine said, as she gave Marc a kiss on the cheek. 'Now I'm sorry to be curt, but I have another important meeting in less than an hour. I'll arrange for someone to visit the orphanage in the next few days to take photographs for false identity documents. He'll also provide you with details of a safe house you can use if things go wrong.'

'Is there anything I can do while I'm waiting?' Marc asked.

'A lot of these soldier types think they know better than the civilians who are looking after them,' Maxine said. 'So try making sure your Canadians don't get bored and try anything stupid. We had a couple of idiot British pilots who started walking around a village in broad daylight. They thought escaping was a big joke. In the end I gave the order to have both of them strangled.'

'You're kidding?'

Maxine shook her head. 'Escape lines require networks of couriers and safe-house owners to help men move south and across to Spain. We'd rather strangle two idiots who can't follow simple instructions than risk the Gestapo arresting, torturing and executing everyone they'd encountered during their escape.'

'I'll tell my Canadians that story if they get out of line,' Marc said, managing a slight smile.

Maxine glanced at her watch. 'Now I *really* have to go.'

'Don't you want the details of the Canadians?'

'Right, right,' Maxine said absent-mindedly. 'I'll send a waiter called John-Paul up. Give everything to him. And they'll serve you a good meal downstairs, no ration tickets or money required. There's too many checkpoints and searches in Paris for my liking, so once you've eaten get the first train back to Beauvais and keep your head down.'

Maxine had the office door open and Marc had to yell after her down the stairs.

'You keep safe,' he said, but she was in a mad hurry, and Marc doubted she'd even heard him.

CHAPTER THIRTY

Eight days later

The nuns found space for the Canadians in a disused room at the back of the convent. If there was a Nazi raid, the pair could drop from a first-floor window and run to a hiding spot in the woods. They ventilated their room so there would be no tell-tale smell, and kept belongings in kitbags so they could exit fast, leaving no trace behind.

Marc would have suggested these precautions, but hadn't needed to. It was clear Noah and Joseph were intelligent and resourceful from the way they'd managed to escape, but Marc had also noted their dark uniforms and the explosive detonators slotted into their ammunition belts when they'd first arrived.

They were more than regular soldiers, but it was never good to know more than you needed to, so Marc hadn't asked what the Canadian's mission had been, and the Canadians hadn't told him.

Marc gave a coded triple-knock before entering the Canadians' room. Noah was reading one of the books Marc had been given in Paris, while Joseph fancied himself as a handyman. The nuns found a regular supply of items that needed fixing and he was currently using wooden batons to reinforce the badly cracked frame of a cot.

'What's this, a half-day?' Noah asked cheerfully, as he eyed Marc's sweaty hair and grubby farm clothes.

'Lunchtime,' Marc said.

'Isn't that usually spent canoodling with the farmer's daughter?' Joseph asked.

'She must find your sweat and cow-manure perfume irresistible,' Noah added.

'Jae works on the farm too, so we smell about the same,' Marc said. 'And how come you're so knowledgeable about my love life? You're not even supposed to leave this room.'

'Young Sister Peter is a proper gossip,' Noah explained. 'Tells us all kinds of things. Not to mention she's got the *sexiest* little bum.'

Marc burst out laughing. 'You can't fancy nuns. They're brides of Christ.'

Noah looked confused. 'What's that supposed to mean?'

'It means, she's married to God,' Joseph explained. 'So according to Catholic doctrine, if you start carrying on with a nun, when you reach heaven you've got to explain to God why you've been screwing one of his wives.'

Marc liked both the Canadians, but they couldn't go out in case they were seen by one of the little orphans. They were usually bored out of their heads and came out with all kinds of random stuff. But before this conversation got too crazy, Marc lobbed a brown envelope into Noah's lap.

'I gave up my lunchtime snog to bring you that,' he said.

Noah's face lit when he pulled two sets of identity documents out of the envelope.

'Beauties,' he said, as he threw the other set at Joseph. 'I think those photos Sister Peter took have captured my essential devastating beauty.'

'They look real,' Joseph said as he held his new French identity card up to the window and studied the watermarked paper.

'They probably are real,' Marc said. 'I'd bet it's easier to steal blank documents than fake them. Besides, your problem isn't gonna be the quality of your documents. Those weird Canadian accents will give you away long before that.'

'I didn't know you were coming with us, mind,'

Joseph said suspiciously. 'Something you're not telling us, Marc?'

Marc was stunned as Joseph held out identity documents and an Ausweiss with Marc's photograph and a false name on them. The Ausweiss was needed to cross the border between German-occupied France and Vichy France.

'I have no idea what that's all about,' Marc said. 'A guy I've never seen before threw the envelope at me and rode off on his bike.'

It was true that Marc didn't know the envelope contained documents for his own escape, but a lie that he had no idea why it had happened.

Maxine had naturally assumed that Marc wanted to complete his escape and return to Britain, but he was crazy about Jae and the documents were a horrible reminder that he might soon have to leave her behind.

'Marc, you look like you just swallowed a dog turd,' Noah said.

'Are you OK?' Joseph added.

'Yeah,' Marc said. He shiftily combed dirty fingers through his hair before changing the subject. 'You're safer now you've got documents and I've gotta get back to work.'

*

Marc and Jae had a regular spot where they met for lunch, close to the lake at the back of Morel's land. She had fresh-picked strawberries, but didn't look happy. The orphanage was a ten-minute sprint and Marc was breathless after doing a return trip.

'Urgent errand for one of the nuns,' Marc said breathlessly. '*Really* sorry. I was gonna say, but I couldn't find you all morning.'

Marc hadn't told Jae about the Canadians. A kiss made up for his lateness and they mucked about, feeding one another the strawberries and trying to stuff grass down the back of each other's shirts.

'I wish we didn't have to go back to work,' Jae said longingly, as she stood up, arching her back and stretching into a big yawn.

Marc was on the point of copying Jae's yawn, but was stunned by the sight of German soldiers running across Morel's fields and an unfamiliar truck ploughing through wheat towards his grain silo. Then he recognised a bald head and a ridiculous swastika armband worn by former orphanage director Tomas.

'It's the Requisition Authority,' Jae said. 'They've come to check on my dad's grain stocks. I'll run round and warn Felix and my dad they're coming.'

She belted off before giving Marc any chance to argue. He followed for twenty metres, but what he saw next froze him in his tracks: his life-long nemesis had emerged from the back of the Requisition Authority truck.

Lanier hadn't stooped to the level of Tomas with

the swastika armband, but he wore the Requisition Authority's pale brown shirt. This was peculiar because as far as Marc knew Lanier worked at a bakery in Beauvais.

He was torn between asking Lanier what he was playing at, and following Jae. In the end, curiosity lost out to protecting his girlfriend.

Despite being knackered after the orphanage run, Marc caught up with Jae as she ran through the main entrance gate of Morel's farm. A gravel path led up to the Morel family home and on to the stable block and farm office beyond it.

Jae had hoped to warn everyone about the Requisition Authority inspection, but it seemed the truck arriving at the grain silo had not been the first stage of the operation. Felix the farm manager, the Morels' three household servants and a dozen farm labourers had already been lined up in front of the stables.

Two elderly German soldiers guarded them in a fairly half-hearted way, while Morel himself argued furiously with the most senior German officer present. The gist of Morel's argument was that the area around Beauvais was under Luftwaffe jurisdiction and that the soldiers had no right to interfere.

In return, the army officer claimed he was working under the supervision of the Requisition Authority, which was a civilian organisation that didn't need

Luftwaffe permission to act.

Jae approached her father and stood close by. He was angry, but still found a second to reassure his daughter.

'Don't worry, sweetheart. It's just bureaucracy. Where on earth were you?'

While Jae and Marc had skirted around the edge of the farm, Tomas, Lanier and a couple of other Requisition Authority officials had walked a direct route from the grain silo and were just arriving on the scene.

'Oh, where *was* Jae?' Lanier asked loudly, as he pointed at Marc. 'Christ, Morel. You must be the only person around here who *doesn't* know he's been at it with your daughter down by the pond every lunchtime for the best part of two months.'

CHAPTER THIRTY-ONE

Everyone was separated and questioned, while the Requisition Authority photographed Morel's grain silo. Morel and Felix the farm manager were taken away in the back of a truck. Marc watched anxiously as soldiers ordered Jae to lead them inside the house so they could search her father's study.

Marc was among the last to be taken into Felix's office for questioning. A pug-nosed Requisition Authority official scanned his identity papers.

'How long have you worked for Morel?' the official began.

'Couple of months,' Marc said.

The official's next statement was spoken rigidly, as if he'd already used the exact phrase many times before.

'If you are aware of any black-market sales of food, or

other illegal activity practised by your employer and you provide us with full and honest details, we guarantee leniency. However, any lies you tell will be regarded as anti-German activity and subject to the full force of the law. Is that clear?'

'I'm just a farm boy,' Marc said.

He'd dealt with far scarier characters than this fellow and didn't feel intimidated. But Marc worried that the effect on some of the other farm hands would be greater. They were simple men, who had families to protect.

'Is that clear?' the official repeated.

Marc nodded. 'I've never been involved in anything, sir. I'm the youngest on the farm, and the most junior. I do what I'm told.'

This was true, apart from the time he'd ridden the cart of black-market food into Beauvais when the usual driver's son had been sick.

'You may be called back for further questioning. You are not to travel more than five kilometres from this area within the next fourteen days without our permission. And if you *should* happen to remember any of Morel's activities that might be of interest, you'll be treated most favourably.'

Marc was worried about Jae, but the Requisition Authority team was swarming over Morel's house and land and he was given no option but to head out of the farm gate.

His head was full of questions as he walked slowly towards the orphanage. Was Morel going to prison? If he did, would he still have a job? And where would Jae go? If she stayed, would Morel let her go anywhere near him?

*

Marc was back at the orphanage before three and crept up to his attic bunk, in case the nuns found a job for him. Victor and Jacques offered a hunting trip, but he wanted to be around when Lanier got back from work.

With all the bigger boys sent away for factory work, Lanier was one of the oldest left in the orphanage. Until Marc's return Lanier had gained a dominant role in the attic dormitory, though he was more arch manipulator than outright bully: the kind of kid who'd steal your school books and blame someone else, or grass you up to your girlfriend's father.

Lanier was also extremely cocky, and Marc hoped he could use this to find out as much as possible about what had been going on.

When he saw Lanier strutting through the orphanage gate, with his new Requisition Authority shirt hanging out of his trousers, Marc moved to a bunk near the door and made sure he looked as miserable as he felt.

'All right, son?' Lanier said brightly, as he strode in with his chest puffed out. 'Good day at the farm? Or maybe not!'

It took a degree of self-control not to plant a fist in

Lanier's face, but Marc needed information more than revenge.

'I thought you worked in the baker's in Beauvais,' Marc said.

'I did until two days ago,' Lanier said. 'But those slave-drivers wanted me in at five every morning, which meant leaving here at four. The owner turned me down when I asked for a raise, then threatened to sack me and tell the nuns I was lazy. That was a big mistake, because I *knew* that fat bastard was getting half his flour illegally from Morel.

'So I popped over to see Tomas at the Requisition Authority and offered my services, with a little information as a sweetener. I get more money, better hours and best of all it's a government job, so they can't send me to the factories, or ship me off to Germany when I turn fifteen in a few months' time.'

Marc snorted. 'And you get to betray your country at no extra charge. Shipping everything France makes and grows off to the Nazis, while our people queue up for stale bread and mouldy cheese.'

'What did my country ever do for me?' Lanier said, then gave a *don't care* shrug before cracking a big smile. 'You're just ticked off because I messed up your love life.'

'At least I've got a love life,' Marc said, as he tutted. 'And what do you gain out of screwing me over?'

Lanier laughed. 'Well, the look on your and Morel's

faces was pretty good. But basically, Marc, you've gotta accept that there are winners and losers in the world. I've used my brain to get ahead. You *act* like some big shot, but what did you actually do? Spend two years poncing around in Paris and get dragged back by a cop, looking like you hadn't eaten in about three months.'

Marc wished he could have told Lanier how he'd really spent those two years. As Lanier squeezed through a narrow gap towards his bunk and started unbuttoning his shirt, he couldn't resist having another dig.

'Reckon I've done you a favour anyway, mate,' Lanier said. 'That Jae's one stuck-up little bitch . . .'

'Oh, shut up,' Marc said.

'My old job at the baker's might be going if you want it,' Lanier teased. 'It's not in my league, but it pays better than Morel's farm.'

Marc didn't rise to this one, so Lanier had another go as he sat on his bed, switching his work trousers for a pair of tattered knee-length shorts.

'Jae ended up at the Requisition Authority office in Beauvais, bawling alongside her stepmother. I tell you, they put on quite a pompous little duet of foot stamping!'

Marc and Lanier had been alone in the attic bedroom, but now a couple of eight-year-olds came in. They looked all sheepish, having just been yelled at by one of the sisters for playing on the stairs.

'And why are you even with a girl like that?' Lanier continued. 'Jae might be into you now, while they're desperate for bodies to shovel shit on their farm, but do you *really* think she'll be interested in someone like you when the war ends?'

Marc finally took the bait and sat up sharply. 'Lanier, *shut* your mouth.'

Lanier laughed. 'Oooh, did I touch a nerve?

'Life's about more than money,' Marc said. 'Now I'm *telling* you to shut up.'

Lanier had to head downstairs to get his dinner. 'Just don't keep us all awake, crying yourself to sleep when Jae dumps you,' he said, as he passed Marc's bed.

Marc's strategy of using Lanier's cockiness to dig out the truth had worked well, but Marc was in a state over the situation with Jae and he'd not anticipated Lanier's taunts making him so angry.

As Lanier swept past the bed, Marc stuck his leg out and Lanier stumbled. He kept upright, but his head glanced off the upright of the nearest bed. It didn't hurt much, but the two eight-year-olds were highly amused.

'You want a fight?' Lanier asked, bunching his fists.

Marc and Lanier had fought dozens of times over the years and the score was pretty even. Part of Marc's brain was saying *back down, don't be crazy*, but he was stressed out and on some primitive level he really fancied trading punches.

'In the back field, right now,' Marc said. 'Unless you're chicken.'

'I'm not scared of you,' Lanier said.

Boys used the back field for fighting because it was behind the burned-out barn, well out of sight of the orphanage and convent.

Lanier looked behind suspiciously as he led the way downstairs, half expecting a shove in the back. By the time they vaulted the wall at the back of the orphanage, word was spreading that there was going to be a big fight.

The field had knee-high grass. The sun was dipping and the ground was covered with wood splinters and chunks of brick, where a German dive bomber had crashed into the barn two years earlier.

Lanier threw the first blow. Marc dodged.

As an audience of younger boys clambered over the wall, Lanier kept punching and Marc kept teasing. The onlookers thought Marc was scared, but he was tiring out an over-aggressive opponent, exactly as Instructor Takada had shown him on CHERUB campus.

'Are you fighting or running?' Lanier spat.

His next punch was high and Marc used the opening, smashing his fist into Lanier's nose. As the crowd gasped, Marc threw a left right combo and Lanier was flat on his back in the long grass.

'Who's brave now?' Marc shouted, as he spat in Lanier's face, then backed off to let him stand up.

The instant Lanier was on his feet, Marc launched a vicious kick. His bare heel connected with Lanier's stomach. As Lanier doubled over, Marc brought his knee up, smashing his nose for a second time.

There were gasps from the crowd as Marc grabbed Lanier by the throat, throttling him as he drove him back several metres and slammed him hard against the charred wooden side of the barn. When Marc let Lanier's neck go, all he could do was snort blood and hold his arms weakly over his face.

Orphanage fights were usually accompanied by cheers and jeers, but Marc's ruthless display of combat skills had stunned sixty boys into silence.

Lanier was completely at Marc's mercy and he had a menu of techniques he'd learned on CHERUB campus. He could punch Lanier unconscious, twist his arm up behind his back and break it, smash a palm under his chin and shatter his jaw, put him in a headlock and snap his neck to paralyse him for life.

Marc hadn't admitted it to himself upstairs, but he'd known he had the skills to win. It was Marc's way of showing Lanier that something significant *had* happened in the two years since he'd run off, but he now had no desire to finish Lanier off.

This fight was no fairer than if he'd picked on one of the eight-year-olds, and Marc suddenly hated himself. The moment when his knee crunched Lanier's nose had felt beautiful, but that made him no better than people like Alain, Fischer and Tomas when they'd beaten him.

Marc was confused and tearful as he backed away from Lanier. Sisters Peter and Madeline had noticed the train of boys heading into the back field and the younger lads dived for cover as the nuns ran around the side of the barn.

'What in the name of God?' Sister Madeline shouted.

She saw Lanier propped against the barn, half unconscious, with bloody hands and face. Marc was three paces further back, sobbing because he felt like he'd turned into the kind of bully he'd always sworn he'd never become.

CHAPTER THIRTY-TWO

Sister Raphael was old school. She worked every hour keeping orphans fed, washed and orderly, but didn't much hold with touchy feely stuff. When she saw Marc was still upset half an hour after the fight, she made an indignant grunt and told him to pull himself together.

Boys who broke orphanage rules got extra chores, or a thrashing. Marc felt ridiculous baring his bum and bending over a desk for a chubby nun, fifteen centimetres shorter than he was. The sting of the metal-tipped cane felt like a visit from an old friend, and when he had half a dozen red welts across his buttocks, Marc got told to pull up his trousers and stop the ridiculous snivelling.

Marc sat on the front steps of the orphanage in moonlight. The hard stone after a thrashing was painful, but he didn't budge because he felt he deserved it as punishment for beating up Lanier.

Every now and then Marc heard Lanier moan as Sister Madeline treated his injuries in the medical room under the stairs. When he saw a slim figure walking a bike up the front path, Marc thought he'd started hallucinating.

'Hey, you,' Jae said softly.

She'd changed out of her farm overalls to go into town. She looked like the Jae Marc knew before he'd left, in a summer dress, cardigan and smart leather sandals. Only the dirt packed under broken nails gave the game away.

'I thought I'd have the devil's job getting inside to speak to you,' Jae said. 'And here you are, right on the front step.'

She sat next to Marc on the step and rested her head on his shoulder.

'Where's your dad?' Marc asked.

'Home,' Jae said. 'One of the Luftwaffe officers who lived with us got him and Felix released. They've emptied our grain silos, and shut the bakery in Beauvais down. Daddy will probably have to go to court. He's brushing it off, but I can tell he's worried.'

'I beat the shit out of Lanier,' Marc confessed.

'Good,' Jae said resolutely. 'I *hope* he's in a lot of pain.'

Marc shook his head. 'I laid into him and I enjoyed it. All my life bullies have thrashed me or beaten me up. I don't *ever* want to act like that.'

Jae put her arm around Marc's back.

'There's evil in us all,' she said. 'When I was little, I used to play with my older brothers down at the pond. A baby duck got separated from its mum. My brother caught it in a bucket and planned to look after it, but he beat me in a board game. So I threw his duck out of the house and let my other brother take the blame for it.'

Marc laughed uneasily. An imperfect world seemed to matter less if Jae was around.

'Did your dad say anything about me?' Marc asked. 'If I turn up at the farm tomorrow morning, he's not gonna come chasing after me with a shotgun or anything?'

Jae thinned her lips and looked determined. 'He's got other things on his mind right now. But I'm growing up and he's short of labourers on the farm. So Daddy can like it or lump it, but I am going to carry on seeing you.'

*

After Jae left, Marc realised he was *madly* in love with Jae and found it scary that someone else had such power over his emotions.

Had Marc punched Lanier out, the other orphans would have regarded him as a conquering hero, or at least showed the kind of reverence his physical dominance deserved. But beating Lanier senseless, then spending three hours snivelling and staring into space had earned Marc loony status instead. Even his ultra-loyal bunkmate Jacques didn't know what to say.

The next morning on Morel's farm passed as normal, but at lunchtime Jae told Marc to have lunch with her at the main house, rather than hiding out by the pond.

'Will I live?' Marc asked warily.

Jae shrugged. 'Daddy's eyes didn't bulge *much* when he asked me to fetch you.'

After they'd stripped wellies and washed hands in the boot room, the Morels' cook served Marc and Jae chicken and wild mushrooms, cooked in wine and served with boiled potatoes.

Marc was touched, realising that Jae had given up a meal this good every time she'd gone to the pond with him. But he expected her dad at any second and found himself apologising to the cook, because his nerves meant he could hardly swallow a mouthful.

Marc thought he'd escaped the encounter, but just before 2 p.m. Morel walked into the servants' dining room. The incident with the Requisition Authority had knocked something out of him. His hair looked flat and he lacked the casual authority that usually made him so intimidating.

The cook gave Morel a plate, but he struck an eccentric pose, eating quickly with a large serving spoon while standing over the kitchen counter.

'Do you read and write well?' Morel asked.

Marc found Morel's tone offensive, as if being an

orphan meant he was ignorant.

'I speak, read and write well in French and German, sir,' Marc said, as the small amount of lunch he'd managed to eat did back flips in his stomach.

Morel smiled slightly. 'German. The language of our future, perhaps?'

'Hope not,' Marc said, although he regretted it instantly because Morel was close to the Germans who lived in his house. But on the other hand, wasn't the German-controlled Requisition Authority prosecuting him?

Marc tried to stay calm and stop over-thinking.

'I bloody well hope to see the back of the Boche too,' Morel said. 'They've left a job for you. Come upstairs.'

Marc and Jae both stood, but Morel made a *down-down* gesture at his daughter.

'You go back to the fields. I only need him.'

Marc looked anxiously at Jae. He remembered what she'd said the night before about her father having to get used to him being part of her life, but her determination did him no good if there were a couple of burly farm hands waiting upstairs with horse whips.

Marc sized Morel up as he led the way out of the basement. He was slim like his daughter, average height. But with Morel it had always been his money and status that was intimidating, not his physique.

'I'm not well liked by my staff,' Morel said. 'Do you

think it likely any of them will denounce me to the Requisition Authority?'

The farm workers constantly slagged Morel off for his tight pay, and reluctance to invest in farm equipment, while no expense was spared on the lavish family house. But Marc sensed this wasn't the time for brutal honesty.

'Everyone moans and groans,' Marc said. 'I don't think there are many fans of the Germans out there.'

'I *could* go to prison for selling my grain direct to the bakery,' Morel said, as they reached the ground floor, and turned on to the much grander staircase leading up to the first. 'Felix and I would probably have stayed behind bars if I didn't have friends high up in the Luftwaffe. They're doing what they can, but Tomas has always disliked me, so I expect he'll keep baying for blood.'

Marc nodded sympathetically. 'I've heard of people deported to Germany as tobacco smugglers, just because they were non-smokers taking a few packs of cigarettes to relatives.'

'The officers who board here tell me it's not much better for Germans,' Morel said sourly. 'German men go into the military, the women into factories, while their children are indoctrinated at school. This entire continent has been enslaved by a tiny cult of brainless thugs and racists.'

Morel stopped on the first-floor balcony and looked at

Marc, to see if he'd understood what had been said. Marc realised he was being sized up and tried to think of an intelligent answer.

'I just hope the Russians can hold out until winter bogs the war in the east down again. By spring, the Americans should be much better prepared.'

'So you believe the Allies can win?' Morel asked, as he approached a set of ornate double doors. 'Or just hope?'

'Believe,' Marc said firmly. 'I'm less sure how much of France will be left standing by the time everyone else finishes fighting over it.'

Morel was clearly satisfied by Marc's answer and laughed as he opened the double doors.

'Speaking of destruction,' he said.

Marc had never been beyond the main hall and servants' area of the Morel house. He was stunned to find himself in a library which ran the entire length of the first floor. There was also a collection of curiosities: scientific instruments, fossils, a huge wooden globe.

The Requisition Authority had been through like a tornado, pulling thousands of books from shelves until it was impossible to move along the aisles. It was a shocking scene, but Marc was mainly just relieved that he wasn't being dragged out to the stables for a whipping.

'It should take you two or three days to put everything back in place,' Morel said. 'Work quickly, or I'll be most displeased.'

'Yes, sir,' Marc said.

He jolted when Morel's fingers dug into his shoulder. 'And the matter of my daughter,' Morel said chillingly.

Marc felt like as if every drip of water had suddenly been sucked from his body.

'I was your age once,' Morel began. 'Fathers protect daughters because they know *exactly* what races through the mind of a teenage boy. It's why we get so uncomfortable when we see some randy young bull like you, with eyeballs wandering up his daughter's legs.

'Marc, I love my daughter more than anything else in the world. I can see how much Jae likes you and if I broke you up it would only drive a wedge between us. But if you ever treat her with anything other than absolute respect, I guarantee you a thrashing that will make the worst Director Tomas gave you look like a pillow fight between two six-year-olds.'

'I'd never do anything to hurt Jae,' Marc said, as he placed his hand over his heart. 'I swear on my life.'

CHAPTER THIRTY-THREE

Two days later

'Daddy misses my brothers,' Jae explained, as she stood with Marc in the boot room at the back of her house. 'I think he actually rather likes having you around.'

Marc had spent the day reshelving library books, and felt guilty because he was clean and dry, while Jae had spent a stormy day in the fields.

'I don't get why you're outdoors while I'm warm and dry in the library,' Marc said. 'Mind you, I'll be finished by tomorrow lunchtime.'

Jae got the cutest look of frustration on her face when she tried pulling her muddy boot off, and Marc stood behind and helped, before nuzzling the back of her neck.

'You could stay for dinner,' Jae said.

'It's my night for washing pots at the orphanage,' Marc

said. 'Best not to piss off the nuns.'

Marc missed Jae as soon as he'd left the Morel's house. It had just stopped raining. The air felt fresh and the first leaves were dropping.

The orphanage was a melee of cooking smells and nuns yelling at kids who'd got muddy playing on the wet grass. Sister Peter whispered, as Marc unlaced his boots on the front step.

'There's a lady waiting at the convent house,' she said. 'With Joseph and Noah.'

Maxine embraced Marc warmly when he reached the Canadians' room. One of the beds had been tilted on to its side to make space. The floor was spread with maps and aerial surveillance photographs of a German airfield.

'I thought you were hiding something from us,' Joseph told Marc. 'But not something *this* big.'

Clearly Maxine had told the Canadians that he was a trained agent.

Marc smiled awkwardly. 'In which case, I guess it's time *I* knew who you two are.'

'We're commandos,' Noah said. 'If we'd got off the beach at Dieppe before getting captured, our job would have been to break away from the main force and destroy a factory near Rouen that makes cockpit instruments for aeroplanes.'

Marc looked around at all the maps, then up at

Maxine. 'Looks like you've found another job for them.'

Maxine nodded and picked one of the photos off the floor. 'It's a stroke of luck, having an agent and two highly capable commandos turning up within a few kilometres of Luftwaffe headquarters. This is your target. It's a Junkers 88 night fighter, equipped with the latest mark four radar.'

Marc looked at the twin-engined fighter. It seemed normal, but had an all-black paint job and a criss-crossed scaffold of aerials protruding from the nose.

'If we can get the radar from a German night fighter, the boffins hope to be able to develop countermeasures against it,' Maxine explained. 'The night fighter radar signals are monitored from the ground. The women who operate the sets tell the pilots what height and speed to fly at, even down to the split second when they need to open fire.

'The RAF have tried jamming German ground-to-air communications, but the system remains stubbornly effective. For every hundred Allied bombers crossing French airspace, we're losing two to the night fighters. That may not sound huge, but it means the average British bomber crew flying three missions a week is going to be taken down by night fighters in under four months. And of course, many other things that can go wrong in a bomber – mechanical failure, flak, or anti-aircraft guns on the ground. At present, the average life expectancy of a British bomber crew is less than two months.'

'Why has nobody ever shot one of these night fighters down and picked up the pieces?' Marc asked.

'It's a ground-controlled radar system,' Maxine said. 'These planes only ever fly over German-occupied territory.'

'So our job is to raid the workshop where the radar sets are fixed?' Marc asked.

The Canadians both laughed, before Maxine explained.

'The radar set is large, and most of the critical components are bolted into the nose of the aircraft. The only way to get hold of an entire functioning system is to steal a plane and fly it home.'

Marc raised one eyebrow suspiciously as he looked at the Canadians. 'You're not pilots as well, are you?'

'They're not,' Maxine said. 'But the British captured several examples of the bomber version of the Junkers 88 in the Libyan Desert, and these have been flown extensively by British test pilots. One of these test pilots will be parachuted in, making you a four-man team.

'A local resistance group has been observing one of the night fighter bases about fifteen kilometres from here for several months. They've given us excellent information on base procedures and security.

'Your job will be to penetrate the base during the chaos of a night-time operation, board a fully fuelled night fighter and fly it across the Channel to an airfield in Britain. The JU-88 has two seats behind the pilot and a spot for a rear gunner lying flat directly below the cockpit, so you all get a ride home into the bargain.'

'Provided we make it off the ground alive,' Noah said.

'Or don't get shot down by a British fighter when we turn up over the Channel in a plane that looks nearly identical to a Junkers 88 bomber,' Joseph added.

'There will be more operational details, of course,' Maxine said.

Marc backed up to the wall, feeling sick. 'I just don't know,' he mumbled.

'Is something the matter?' Maxine asked anxiously.

Marc hesitated before answering. 'It's just, this feels like my home again now. I've got friends, and . . .'

Maxine looked confused. 'You've got friends on CHERUB campus too: PT, Paul, Rosie, Joel. But this is what you've trained for. I set all this up, assuming you were raring to get back to campus.'

Joseph smiled at Maxine, then put an unwelcome arm around Marc's back. 'Don't you know the young fella's in love? Couple of nights back he half killed some lad just for making a few sly remarks about her.'

'Ahh . . .' Maxine said, as her lower jaw dropped.

She didn't know how to react. As an experienced resistance leader, Maxine expected everyone who signed up to follow orders. But Marc was hardly a normal case. He'd signed up when he was twelve years old. He'd already shown bravery on two daring missions, escaped captivity in Germany and was still only fourteen.

'If that's what you want,' Maxine said weakly. 'I suppose I could make up the team with one of our other people up from Paris. But it's very short notice.'

Marc felt bad: not just for letting down Maxine, but also the idea of abandoning Henderson's team. Rosie, Paul, Joel and PT *were* the strongest group of friends he'd ever made. But he loved Jae so much that the thought of leaving her made him feel like his body was being ripped in half.

While Maxine and Marc didn't know what to do, Noah faced Marc off. He was an intimidating man. His enormous chest came level with Marc's face and his arms looked like they could punch you clean through a brick wall into the next room.

Marc backed up when Noah's hand came out of his trouser pocket, half expecting a knife or knuckleduster. Maxine looked anxious too, but it was only a photo. One of the few colour photographs Marc had ever seen.

'That's my wife and twin daughters,' Noah said. 'I could get court-martialled for taking their picture with me on a covert operation, but this photo is all I've ever seen of my babies and I love my wife.

'I sit here all day thinking about 'em. But every one of us Canadians are volunteers. I came over to fight, because what kind of world would my wife and daughters end up living in if men like me sit on their asses?'

After the cruelty Marc had seen in Germany, he realised he couldn't leave other people to fight the war just because he'd fallen for a girl. He felt a touch pathetic as he looked away from Noah towards Maxine.

'If it needs doing, then,' Marc said determinedly. 'When does this happen?'

CHAPTER THIRTY-FOUR

Two days after finding out about the mission, Marc met Jae for lunch by the pond. For security purposes it would have been better to just vanish, but Marc cared too much to simply abandon his girlfriend without a word.

Marc explained how he'd been trained in Britain, arrested in Lorient, spent a year in Frankfurt, escaped and wound up back at the orphanage more or less by accident. It wasn't an everyday story. Jae had trouble believing all of it and once she did she sounded both shocked and impressed.

'Now I have to leave,' Marc said.

'When?'

'After work, provided the weather stays clear.'

'What's the weather got to do with it?'

Marc shouldn't have told her, but hiding stuff from

someone he was so crazy about seemed wrong.

'Parachute drop,' he explained. 'The pilot can't accurately drop the equipment for our mission if it's raining, or there's low cloud.'

'It's so soon,' Jae said. 'I can hardly take this all in.'

'I know it's brutal, but that's how it always works, for security. I only found out the day before yesterday. The fewer people who hear about an operation and the less time there is before it kicks off, the less chance there is of someone snitching, or getting picked up and interrogated.'

'I want to come with you,' Jae said, close to tears. 'If my dad gets sent to prison I'll be all on my own.'

'Your dad's a wily one,' Marc said. 'He'll work out a deal. Wouldn't be surprised if he stitched Tomas up into the bargain.'

'I hope,' Jae said.

'Besides, when you tell your dad that I've run away again, he'll be able to say *I told you so*.'

Jae managed a slight smile. '*Why* does this have to happen?' she moaned.

'I wish it didn't,' Marc said. 'But the war's too important to sit out.'

'We might never see each other again,' Jae said.

Marc's eyes glazed over as he shook his head. 'Don't say *never*. We'll be sitting here again before you know it.'

Marc pulled a knife out of a leather sheath. Jae looked

curious as Marc stroked her long hair.

'Memento,' Marc explained. 'OK?'

Jae nodded, then sniffled as Marc sliced off a few dozen long strands of her hair. He tied the bunch in a knot and gave them a kiss before pushing them into his pocket.

'What do I get to chop off?' Jae asked, as she ran her hand over Marc's head.

Marc had been virtually bald when he got back from Germany, and even now there were only a couple of centimetres of hair on his head.

'I want your shirt,' Jae said, as she undid the top button.

Marc looked at her like she was mad. 'I've been wearing this all week. It stinks.'

'Of you,' Jae said. 'I like your smell.'

'I'm kinda wearing it,' Marc pointed out.

Jae thought for a couple of seconds. 'Put it on the wall outside the orphanage when you leave. I'll ride over and collect it before sunrise.'

The idea seemed both romantic and crazy, but it showed Marc that Jae really cared about him. They were both miserable so he tried a joke.

'Expect I'll change my socks as well before I set off,' he said. 'Do you want them too?'

'Oh, you're funny,' Jae said drily, before rolling into Marc's lap so that he could give her another kiss.

Then she dug her nails into his wrists and begged him not to leave.

'I've got to,' Marc said, as he tasted Jae's salty tear on his tongue. 'Please don't make this harder than it is already.'

*

The first time Marc left the orphanage he'd been running, scared out of his wits. This time he was trained, he was in love, he had a mission. He didn't feel like a boy any more, though trailing through countryside struggling to keep pace with two burly Canadian commandos was a stark reminder that he wasn't a man yet, either.

It was after curfew so they kept to the woods, navigating by compass and moonlight. Four kilometres east of the orphanage they met up with a guide from a local resistance cell.

'Nice evening for hunting,' the guide said.

Joseph answered with a pre-planned reply. 'The only thing I've caught is a cold.'

The woman was in the early stages of pregnancy, dressed in trousers and rope-soled shoes. For security's sake, nobody exchanged names, or details of where they'd come from.

Despite the woman's state, the pace didn't flag. Marc battled a stitch down his right side, but with a pregnant woman leading the way he was too proud to admit anything beyond a stiff back from labouring on Morel's farm.

With half of Europe to run and their best men committed to the brutal war with Russia, German security was spread thin. The guide led them through woodland, across open fields and even braved a section of country road. Despite a full moon, they didn't spot a German patrol, or even hear a passing vehicle.

Their first halt was just before midnight.

'Welcome,' a man said, before beckoning the quartet down a short ladder. He wore an English huntsman's suit, and his accent established him as true French aristocracy.

They stepped down into an air-raid trench, built within sight of a country house that made the Morels' place look like a tin shack. Marc shook hands with two lads, probably a year either side of his own age. Although nobody gave names or asked unnecessary questions, these were undoubtedly sons of privilege, in hand-tailored sport coats and polished leather riding boots.

Before the war, much of France's wealthy elite had held more sympathy for Hitler than their own socialist government. But two years of Nazi occupation had shattered a lot of illusions, and Marc drew strength from the fact that he was squatting in a muddy shelter, toe-to-toe with some real-life toffs.

'You've got half an hour to rest,' the aristocrat said, as he glanced at a fine wristwatch. 'There's bread and cheese, plus a little wine while we wait for the show to start.'

Marc thought *show* was a reference to a parachute drop. But when they eventually moved from the shelter to a hedge overlooking the rear of the huge house, it seemed something more elaborate was about to take place.

Twenty members of the local resistance were spread over the perfectly mown lawn behind the grand house. Old men, younger women and a few peasant boys. Hay bails had been lit at the corners and there was a distant hum of two aircraft.

Marc was impressed by the scale of whatever was about to take place, but knew better than to ask what it was. As the sound of propellers grew, a pair of peasant girls dashed to the far end of the lawn, close to a lake. They waved phosphorous-dipped sticks that burned bright white.

When the two little planes came into view, they swept close to the tree tops, silhouetted against the full moon.

'Surely that's too low?' Noah said anxiously. 'What are they playing at?'

Marc had done parachute training and realised from their height and approach angle that these planes weren't about to drop anything.

'They were coming in to land,' Marc gasped, as he pointed at the burning hay stacks and the white sticks. 'They've marked out a perfect landing strip.'

'We didn't want to spoil the surprise,' the posh landowner said, as he beamed at the two Canadians. 'Quite a show, isn't it?'

The approaching planes were RAF Lysanders. Their pre-war design was outclassed by newer aircraft, but with wings set high above the fuselage and a sturdy undercarriage the tiny Lysanders could take off and land in rugged spaces barely longer than a football pitch.

The first Lysander came down hard, stopping so abruptly that it pitched forward on to its nose. The locals seemed unfazed as they raced up to the plane, hung off the tail fins to set it right and then picked off lumps of soggy turf clinging to the propeller.

The small cabin door opened. Two passengers stepped out, dragging pallets of equipment behind them. A man and a woman who'd run from the house replaced them on board, while two peasants loaded the plane up with a stack of file boxes.

Rather than let the plane turn under its own power, half a dozen resisters lifted the tail and pivoted it so that it faced back along the field. Less than a minute after landing, the Lysander pilot had throttled up. Twenty seconds after that he was back in the air, skimming pear trees at the end of the makeshift landing strip.

'We've got it down to a fine art now,' the younger of the well-dressed boys told Marc proudly. 'One of our early landings hit a cow. Made a terrible bloody mess.'

The second plane had circled while the first was on the ground and it landed within seconds of its sister leaving the ground.

Joseph glanced at Marc. 'How come we don't get to go home that way?' he asked, only half joking.

Marc smiled. 'I think that's the first-class service. Not for nobodies like us.'

The second plane delivered a single passenger, along with crates of weapons which got dragged out by a swarm of locals. Leather suitcases and a badly injured RAF navigator were put aboard for the return trip. Shortly the plane was aloft, the flaming hay stacks had been doused and local resistance was dispersing into the surrounding fields.

One of the resisters pointed Joseph and Noah out to the Englishman who'd stepped off the second Lysander.

'Ahoy there!' the Englishman called, as he approached the hedge. 'I'll wager you're my chaps.'

He was every bit the stiff-upper-lipped RAF man, with fiery red hair, leather flight jacket and a university scarf around his neck. Apparently the RAF didn't pay the same meticulous attention to detail that the British secret service did when sending men into France undercover. Marc's first thought was that the fellow urgently needed peasant clothes and some dirt under his neatly trimmed nails.

'Squadron Leader Davey,' he announced. 'Bloody nice show, eh? All the packages off the planes are numbered. Our gear is in one, two, five and six. Let's scoop it up, then get this show on the road.'

Noah looked pissed off, and decided to set the Englishman straight over who was in charge.

'We're not *your* chaps,' Noah said, as he placed his vast hand on the Englishman's shoulder. 'When we get up in the air, you're the boss. When your feet are on the ground, *you* do what Joseph and I tell you. And lose that jacket and scarf. You couldn't look more English if you draped a Union Jack over your shoulders.'

'Now you listen here,' Davey said, but tailed off abruptly when Noah growled like a dog about to bite. 'Shut your hole, you stuck-up English bastard.'

Clearly the Squadron Leader wasn't used to burly Canadian soldiers threatening him, but while the Englishman's shocked expression was a treat, Marc knew the team had to work together. He was still thinking of the best way to intervene when their pregnant guide did the job for him.

'Stop waving your dicks at each other,' she said furiously. 'The Germans might have seen the landings. Let's grab our gear and get moving.'

CHAPTER THIRTY-FIVE

There were another ten kilometres of fields and footpaths to walk. Besides Davey the pilot, the Lysanders had delivered guns, ammo, grenades, spent parachutes, wire cutters and a specially built Junkers 88 engine-starter unit.

Noel took the biggest load and didn't even strain. Marc ended up with a light-but-bulky pack of commando clothing on his back, and a shoulder-crippling satchel filled with bullets and grenades. Even the pregnant guide carried her share.

A farmer loaded their equipment on to a cart for the last stretch. It went direct to the stables where they'd spend the day, while the guide took Marc, the Canadians and Davey on a detour along a railway line so they could check out their target.

Noah pulled binoculars and crouched at the tree line as the sun rose. It looked like it had been a quiet night for the Luftwaffe, with three-dozen twin-engined JU-88s parked up neatly, wearing engine covers and camouflage nets. Their hulls were painted matt black so they'd be invisible in the night sky.

Davey spoke in a whisper. 'From what I can see here, there's only a couple of planes with the mark four radar here.'

Noah and Joseph were more concerned with base security: studying entry and exit gates and the wire perimeter.

'I thought you'd like to get a sense of the place,' the guide explained. 'But there are regular patrols, so we can't hang around.'

Back at the railway line, their guide wished them luck, and told them to walk for about a kilometre. The men hugged her in turn.

Davey looked surprised. 'Not staying with us?'

This operation had been set up at short notice, but Davey's naivety still shocked Marc.

'She probably doesn't even know where we're staying, or where the cart with our equipment went,' Marc explained, as they walked along an overgrown single track. 'If she's arrested by the Gestapo, she can't name names that she doesn't know.'

Beyond the Luftwaffe base the single track hadn't

been used in a good while. Sections of rail had even been cut out and sold for scrap.

Their next guide was a chubby lad, no older than ten. His cover for being out in the woods was a snail hunt and as he walked he prevented escapes by flicking down snails climbing the side of his swinging bucket.

The boy's family greeted them, but took care not to exchange names. It was a tumbledown farm, with rotting fences and half the roof missing from the stables. There was a single skinny dairy cow, sheep roaming free and the land divided into tiny patches growing a bit of everything.

These were poor people. The contrast to the grand lawn where the Lysanders touched down couldn't have been greater, but it encouraged Marc to think that people from such contrasting backgrounds were willing to risk their lives to kick out the occupiers.

Marc was shattered from a day's farm work, followed by the night-long trek. He breakfasted on milk, cheese and fresh-baked bread before pulling his boots off, curling up on a bed of straw and falling effortlessly to sleep.

*

It was early afternoon when he woke. The Canadians were snoring, but Davey was awake. He'd been found some more suitably French clothes, and he sat in the sun outside the stable playing a lively card game with the snail-hunting ten-year-old and twin sisters a year or two younger.

After watching for a while through one half-opened eye, Marc stepped over Noah and joined in. As Davey only spoke schoolboy French, Marc found himself in the role of translator. It was hardly best security practice to let a pair of eight-year-olds know there were undercover agents in the stable, but that damage had already been done.

Life on a remote farm wasn't exciting, so the kids were excited to have a real-life pilot in their house. Marc translated as Davey explained everything from the basic principles of flight, to the synchronisation gear that enables you to have a machine gun mounted behind a propeller without shooting off the blades.

Davey also knew card tricks. Marc knew better ones, but his skills were rusty and the three kids howled with laughter when he kept getting them wrong.

Mucking around was more than a way to pass time. It helped Marc keep his mind off the dangerous mission that was now less than ten hours away, and visions of Jae that were almost powerful enough to make him stand up and run back to her.

The mood got more serious when the Canadians woke and the kids were sent off to do chores. Davey had brought copies of base maps drawn by the local resistance group. Plans have a way of deviating off course, but that didn't stop the quartet running through every detail several times.

Their hosts had gone all out on the evening meal. There was a large roast chicken, served with potatoes and carrots tossed in garlic butter, red wine, and a peach and raspberry tart with cream fresh from the cow.

The twins hugged everyone and gave Davey a kiss on both cheeks before walking upstairs to bed. The older boy and his father followed the Canadians outside and started sorting through the equipment that had been brought in on the cart.

To minimise risks to the local civilians, Marc, Davey, Joseph and Noah stripped off their civilian clothes and donned British army boots and dark grey commando gear.

Savage reprisals were one of the ways the Germans tried keeping resistance activity under control. Resistance shoots a German Officer: ten people get taken out of prison and hung in the spot where it happened. Resistance catch a man who sabotaged a tank: a tank battalion turns up and demolishes half of the resister's home village.

According to the new dog-tags and military identification papers that Davey had brought off the Lysander, Marc was now a Paris-born sixteen-year-old who'd joined the Royal Navy as a cabin boy, then been attached to the Royal Marines Commando force because he spoke good French. He didn't look sixteen, but some sixteen-year-olds don't look sixteen either.

Joseph and Noah also had new dog-tags and clean identities. Their real identities had been listed by the Germans when they'd been captured after the Dieppe raid, and that would blow their cover story.

They set off after 8 p.m., with the dipping sun warm on the back of Marc's head. He felt invincible, dressed in black, with a metal helmet, a pistol, a pack filled with ammunition and grenades, plus his trusty throwing knife wedged into his belt.

But doubts set in as he walked. He'd been through a lot over the past two years, but the next few hours looked set to be the most precarious of them all. He kept thinking how it was a full day since he'd last kissed Jae, and imagining a gravestone with the words *died aged fourteen* . . .

CHAPTER THIRTY-SIX

To ensure this was seen as a commando raid, not a resistance operation, the first job of the evening was to jog four kilometres and hide four spent parachutes. Newly dropped agents and commandos were trained to hide their chutes, preferably by burying them or weighing them down and throwing them into deep water.

In reality, few agents had the luxury of enough time to hide their parachutes completely and in this instance the plan was to hide the unfurled parachutes with the express intention that the Germans found them.

The quartet spread out several hundred metres, over an area of farmland which would make a good drop zone for real paratroopers. It also happened to be owned by a resistance-friendly farmer who would *discover* one of the chutes early tomorrow morning, if Luftwaffe search parties didn't get there first.

Marc found an irrigation ditch for his chute, kicked some dirt over it and then weighed down the billowing silk with large stones. For extra effect, he abandoned a parachutist's mini-shovel and a broken British Army compass nearby.

By the time he rejoined the other three, the twilight had burned out and a light drizzle had begun.

Noah was so big he seemed like part of another species, but this oak of a man kept stopping to pull down the back of his trousers and squat in the bushes.

'Always gets the squirts something rotten,' Joseph explained. 'He was the same before Dieppe, and the lads he trained with tell me his parachute jump school days weren't too pretty either.'

Noah's nervy bowels reminded Marc that he ought to have been all clammy hands and thumping heartbeats himself. But he'd been in situations like this before and survived, and although this operation would be more intense than his escape from Germany, he'd been alone then. He drew strength from the three comrades walking alongside him.

They were under a kilometre from the airbase, and Noah was doing his business on the other side of a hedge, when they heard a dog bark. They were on open ground, which made it hard to hear which direction the bark came from, but it was definitely something big and vicious, rather than your friendly neighbourhood pooch.

'Everyone down,' Joseph whispered.

Marc dropped on to one knee and opened the leather popper of his pistol holster.

'Patrol,' Davey whispered, as they heard legs rustling through the grass on the other side of a tall hedge.

Marc wondered where Noah was as the dog barked again. A bored-sounding German told the animal to shut up. Then a different German said, 'Scheisse,' which Marc knew meant shit.

Moments later there was torchlight behind the hedge, followed by a shout of, 'Hands up! This is a secure area!'

Marc knew it was Noah, even before the tips of his huge fingers got silhouetted against the moonlight above the line of the hedge.

'What are you doing here?' one of the Germans asked suspiciously. 'Are you alone?'

Marc heard a squeaking sound. Joseph was screwing a silencer to his pistol and Marc dipped into the pocket of his combat jacket to reach for his own.

As a pilot Davey had done no combat training, so Joseph gave him a hand signal to lie flat before tapping Marc's chest and pointing left, then right.

As Joseph peered through the hedge from directly behind the two Germans, Marc tried getting a view from a few metres further back. The scene was almost comic, with Noah standing bare-assed with his trousers around his ankles and the two Germans facing him, one with a growling Doberman on a leash and the other with his rifle aimed at Noah's head.

Marc looked over towards Joseph, but the moonlight was blocked out by the hedge. The next thing he heard was rustling branches, followed by the double pulse of two silenced pistol shots.

Only a clean head shot guarantees instant death, which is vital when you're shooting a man who's only a trigger pull away from killing your comrade. Joseph had put two expert shots through the brain of the German with the gun.

Marc realised what had happened and took aim at the other German as he turned to run. Joseph's third shot came in the same instant, and it wasn't clear who hit the Luftwaffe security officer between the shoulder blades and who got him a few centimetres higher in the base of the neck.

As the second German fell he released the Doberman. Joseph hadn't anticipated this, and forced his way through a gap in the hedge, only to find himself knocked down by a heavy dog, which promptly sank its teeth into his thigh.

Marc had the gun ready, but didn't dare shoot when a bullet could easily pass through the animal, straight into Joseph. As Joseph fought the powerful dog and Noah pulled up his trousers, Marc forced his way through the hedge and threw his knife.

His movement was restricted by the tangle of leaves, but he managed to hit the Doberman in the back. He'd ruptured something major, because the dog spasmed and a jet of blood shot more than two metres into the air.

With the dog weakened, Joseph managed to prise its jaws off his leg. He then ripped out his own knife, slashing the animal's throat as it made a final desperate squeal.

'Bugger,' Marc gasped, as he glanced about at blood and gore.

There were two dead Germans, one with brains spattered all over the tall grass. The Doberman was dead but twitching. Joseph screwed up his face in pain from the bite, while his clothes squelched from all the dog blood.

'What now?' Noah asked, as he walked up to Joseph. 'You OK, pal?'

'All because of your damned bowels,' Joseph said, half angry, half joking.

It took another minute for Joseph to get back on his feet. Marc tried thinking through the implications of what had happened while he pulled his knife out of the dog and wiped it on the grass.

'I say we throw the Germans and the dog in the ditch we passed a few hundred metres back,' Marc said. 'Then we booby trap the bodies with a grenade and move across to the other side of the base. If we hear a bang, we'll know we've been rumbled.'

Noah nodded, impressed, as Joseph glanced at his watch.

'Works for me,' Joseph said. 'If our timings are right, the British bombers should have the night fighters taking off and the base in a state of chaos within the next half-hour. With any luck, a routine patrol won't be missed much before then.'

'What if it it's a local who discovers the body?' Davey asked. 'They'll get blown up just the same.'

'Could happen,' Marc said. 'But this is a restricted zone. I'm open to any better suggestions.'

Nobody had any, so they dragged the two dead Germans and the Doberman towards the ditch.

Joseph was hobbling badly and Davey strapped his leg with a bandage as Marc and Noah stood in the ditch, piling the corpses on top of each other, then setting up a trip wire so that a grenade would explode three seconds after anyone came near the bodies.

By the time they'd cleaned off as much blood as possible in a horse trough and repositioned on the opposite side of the base, they were expecting the night fighters to be scrambling to take on incoming bombers. But midnight passed and Davey grew dejected when it reached 1 a.m.

'They didn't give me an exact target, but I was told that tonight's raid would be over Germany,' he explained. 'Any bombers that pass over now would be on a suicide mission: they'd never reach Germany and get back across the Channel before daybreak.'

'How can they not come?' Marc asked. 'I thought the RAF was desperate for one of these radar sets.'

'Any one of a million reasons,' Davey explained. 'Bad weather over the target in Germany. Emergency reassignment. Or maybe they got shot-to-hell on their last raid and there's not enough of the poor buggers left to form a squadron.'

Joseph grunted with frustration. 'Your plan said if it didn't happen tonight, we'd dig out our parachutes, go back to the farm and try another day. But they're going to miss that patrol we killed. There'll be a major search through this whole area. Security will be beefed up for weeks, if not months afterwards.'

'Not to mention possible arrests and reprisals against the locals for killing two Germans,' Marc added.

'So we're screwed?' Noah said, shaking his head with frustration.

'I *can* see one option,' Joseph said. 'We needed the British bombing raid to create a distraction on the fighter base. But Noah and myself are demolition experts, carrying backpacks full of explosives and grenades.'

Noah smiled before taking up his young partner's

train of thought. 'If we snuck in and wired up the base fuel tanks, would that make a big enough bang for you?'

Davey scratched his beard. 'We'd be the only aircraft in the sky. The Germans could scramble whole squadrons of aircraft to hunt down one of us.'

'What if we disrupt their communications too?' Marc asked. 'Cut the phone lines, blow the radio masts.'

'That would certainly help,' Davey said. 'If I hedge hop – that's keeping the aircraft low – we'd be below most German radar systems. It could work out, as long as I have time to get up in the air, and get ten minutes flying in before they start looking for me. Radar picks up bomber squadrons easily enough, but I saw last night how one or two Lysanders can sneak in and out of occupied France without too many difficulties. How are you set for explosives?'

'We've got no detonator cord,' Noah said. 'But we've got fuses, grenades and a few sticks of plastic.'

Joseph nodded. 'A little explosive goes a long way, if you put it in the right spot.'

'RAF chaps die every night because of this radar,' Davey said. 'So if you're willing to try making some chaos, I'll do my damndest to get a bird off the ground.'

Davey had got off on the wrong foot with the Canadians, but the experience they'd just shared had brought them closer, and for the first time Joseph gave the RAF test pilot something approaching a look of respect.

'We're agreed then,' Joseph said. 'Noah, get the base plans out. Let's work out who needs to do what.'

CHAPTER THIRTY-SEVEN

Joseph studied plans, divvied up the jobs, made everyone synchronise their watches and set the kick-off time for 1:35 a.m.

The Luftwaffe base remained calm as Marc squatted beside a thick tree trunk eight metres from its perimeter. The night-fighter pilots knew there would be no raid over Germany, but would stay on alert for a couple more hours in case of attacks on French targets.

Marc's task was taking down communications. He watched the second hand on his pocket watch. When it swept past twelve, indicating 1:31, he stood up and pulled out his knife. The base was in a remote area, but the location of a telephone junction box just beyond the security perimeter had been marked on the base map supplied by the local resistance.

After a run of less than ten metres, Marc used the knife to lever the metal cover off the junction box. He then dug it in behind the three wire connections and yanked it forward to rip them out of their sockets.

With the base telephones killed, his next task was dealing with a bank of radio antennas, which lived in a three-storey air traffic control building a few metres inside the perimeter. He switched his knife for a hand grenade and crouched down beside the fence, waiting for a distant bang that would signal the time for his next move.

Dead on 1:33 there was an explosion at the opposite side of the base. This was caused by Squadron Leader Davey triggering the booby-trapped bodies, and lobbing a couple of grenades into the surrounding trees to make sure everyone on the base knew something exciting was going on.

Joseph's theory was that the explosions off-base would set all the armed security patrols running away from the base, and away from the lines of parked Junkers night fighters in particular.

Davey claimed he could run a hundred metres in under ten and a half seconds and seemed confident that he could make it through a hole Noah and Joseph had already made in the perimeter fence before anyone got near the scene of the blast.

Marc had no way of knowing whether this had

worked out as he pushed small pieces of rag into each ear, before hauling himself up the base's wire perimeter and bravely throwing himself on to the coils of barbed wire at the top.

The plan was that a second coat and Davey's leather flying gloves would stop him getting sliced up, and it didn't work out too badly.

Once he was sprawled over the wire barbs, Marc swung one leg inside the perimeter and threw himself at the ground on the other side. He'd hoped to land feet first, but the barbs snagged a belt loop on his trousers and he ended up plunging head first, with only his hands and a metal helmet to save him.

A painful shockwave went from his palms to his shoulders, but he had to ignore the pain and duck because a cluster of excitable Luftwaffe men stood less than twenty metres away. Fortunately they were all studying the plume of smoke rising from the trees at the other end of the airfield.

Climbing up to the roof and destroying the radio aerials on top of the control tower would take too much time, so Joseph had instructed Marc to approach the ground floor of the control building and roll three grenades through the front door.

Marc pulled the pins as he approached, using his teeth for the third one because his hands were full. As all the action in the control tower took place at the top, Marc had hoped to find an empty stairwell behind the door. He actually got a porky German technician who didn't appreciate having three grenades bounced towards him.

Marc shut the door and ran flat out, followed closely by the screaming technician. The narrow building didn't collapse, but the simultaneous blast of three grenades sent a shockwave up the staircase. It blew out every window, including the huge top-floor panes that overlooked the airfield. It also sent thousands of glass and wood shards into the crowd of Germans below.

Marc couldn't see his watch in the dark, but knew he only had seconds before the main event kicked off. He clattered into a German pilot, charging out of a wooden hut.

'Who are you?' he demanded.

The pilot was off balance, with a parachute hooked over one arm and pulling his flying goggles on with his free hand. Marc pushed on, but the pilot grabbed his shoulder and spun him around.

The German looked shocked as he recognised Marc's commando gear, but before he got a proper hold on Marc, the teenager lashed out, catching the side of his head with a flying elbow, then plunging a knife into the pilot's gut as he crashed against the wooden hut.

Marc ran on with the bloody knife clutched tightly. He broke out on to the open airfield tarmac, with the criss-crossed runways lined up ahead and rows of night fighters parked to his left.

Flames crackled and men who'd been sliced by flying wood and glass screamed. Marc wore black and was now far enough away from the control tower to be indistinguishable from all the panicky Germans as he ran towards the parked aircraft. He'd made less than sixty metres when the first of several explosions set by the Canadians ripped through the base's partially-buried fuel store.

A sympathetic fuse meant that the ammunition store erupted a few seconds later. There was screaming and the whole sky lit up. Even with the rags wedged in, Marc found his ears ringing as the searing heat blistered the skin on the back of his neck.

Another hundred metres took him to Davey. As chaos unfolded, the Squadron Leader had located a JU-88 fitted with the Mark IV radar and peeled back the camouflage netting from it, but he was gesticulating wildly with the starter unit.

Marc couldn't hear over roaring flame, crashing debris and ringing ears, but he grasped that Davey needed to adjust something in the cockpit before the engine would start running.

'Have you seen Noah and Joseph?' Marc shouted.

He didn't get an answer. Instead, Marc found himself standing by the propeller with the starter trolley. To save weight, the JU-88 didn't have the batteries and starter motors to start the engine turning. There was no certainty of being able to find a starter unit when they took the aircraft, so Davey had brought one the British had cobbled together to start the JU-88 bombers they'd captured in the Libyan Desert a few months earlier.

'Push the button,' Davey shouted.

Marc's arms sagged as he stood at full reach, holding the starter motor up to a slot in the side of the engine. The blast of the propeller knocked him back as the engine caught and he stumbled clumsily as he ducked under the fuselage to start the engine on the opposite side.

There was still no sign of the Canadians as Marc started the second engine and kicked the starter blocks from under the front wheels.

'Get in,' Davey shouted. 'This plane is our mission. We can't wait for them.'

Marc looked around desperately as he put a boot on the ladder that led up to the cockpit. There was no room for his backpack inside the tiny cockpit, and he threw it off, leaving just his pistol, knife and a few grenades hooked to his belt.

'I'll leave the cockpit open,' Davey said, as Marc settled into a seat directly behind the pilot. 'Get your safety harness on.'

Marc glanced around as he settled into the cockpit. They'd been lucky nobody had noticed as they'd started the plane, but it would be hard not to as Davey throttled up and started taxiing towards the runways.

They were rolling towards the scene of the main fuel explosion. The orange light was blinding and the heat of the flames was intensified by the cockpit glass.

In the confusion, none of the Germans they passed made any attempt to stop them, and with the control tower destroyed, nobody understood what was going on.

'Sighted, three o'clock,' Davey shouted.

Marc didn't hear, but the pilot put the wing flaps up and they slowed so dramatically that Marc feared they'd go nose up, like the Lysander he'd watched one night earlier. He only understood what was happening when they came to a halt and he saw Noah limping towards the plane.

With no ladder, the only way into the cockpit was to duck under the wing and pull yourself up. Noah was strong enough, but Marc had to slide out of his way and was horrified by bright red, circular wounds where most of his hair used to be.

'Joseph's dead,' Noah shouted. 'Let's move.'

Marc tugged a leather strap to pull the cockpit shut, then settled into a tiny flip-down jump seat. As he hunted for the seat harness, Davey turned sharply and headed for the runway.

But while nobody had interfered up to now, half the base had watched a man in British commando uniform climb aboard. Fortunately the armed security patrols were nowhere near, but the same couldn't be said for the base's fire-fighters.

The main fuel fire was beyond the scope of these teams, and they were concentrating on smaller fires causes by flying debris. As Davey turned on to the runway and began accelerating, two powerful water jets swung into their path.

There was nothing in the flight training manual about what you're supposed to do when you're hurtling along a runway at full throttle and someone aims a fire hose at you, but Davey strongly suspected that the pressure of several tons of water hitting the tail would send them dangerously off course.

'Hold on, boys!' he shouted.

Davey swerved off the runway. His first plan was to keep up the pace and swerve back on to the tarmac, but at this speed any dramatic turns would most likely rip off the undercarriage.

The grass up ahead looked clear and flat, though it was impossible to be sure. Davey had to take a split-second choice between throttling back and giving up, or keeping on full throttle and hoping he could reach take-off speed before they hit the base's perimeter fence, or some other unseen obstacle.

The top of Marc's head slammed against the inside of the cockpit as the engines went at full throttle.

Everyone's eyes were stung from the acrid smoke produced by burning aviation fuel.

'I'm sorry, boys,' Davey shouted. 'Brace yourselves; I don't think we've got the speed.'

But the bumps had stopped before the last word was out of Davey's mouth. Marc could see nothing but the flaming airbase reflected bright orange in the cockpit glass as Davey pulled back on the control stick.

There was a crashing sound, as branches thrashed against the undercarriage. The engine made several loud misfires as the plane lurched violently sideways, snagging on something heavy. Davey threw the control stick in the opposite direction, more in hope than expectation, but they broke clear of the trees.

Marc put his head against the seat back as the flames on the ground shrunk from view.

'We're full of fuel,' Davey said happily, as his eyes darted around the cockpit checking the rows of illuminated gauges. 'Oil pressure good. Controls feel OK.'

Marc looked across at Noah whose face was smeared in blood. 'You OK?'

'Just burns, I think,' Noah said. 'We were still too close when the fuel tanks went up. I got hit by shrapnel, but Joseph was closer. The fireball threw him a good thirty metres into the air. I didn't see a body, but there's no way he could have survived that.'

'I'm sorry, mate,' Marc said, as he felt Noah's hand trembling above his own on the narrow armrest between the two seats.

'This one was never going to be easy, was it?' Noah said.

'Quite a ways from home yet, chaps,' Davey said anxiously, as he pressed his face against the cockpit glass to cut out the reflection from the illuminated dials. 'Whatever we snagged on the way up has shredded our right tyre.'

'Can you land on one?' Marc asked.

'No idea,' Davey said. 'Looks like I'll be giving it a try though.'

CHAPTER THIRTY-EIGHT

Ironically, after all Marc had been through it was the combination of a hard seat and the scabs from his thrashing by Sister Raphael that gave him most discomfort on the flight home.

Davey was targeting an RAF airfield at Bexhill, on England's southern coast. It was less than two hundred kilometres from their take-off point, making a forty-minute flight for the night fighter. They flew at two hundred metres. This was below German radar coverage, but also low enough that an unexpected hill or a tiny lapse in Davey's concentration could easily lead to a crash.

The coast was well defended, so Davey went higher and took a slight detour to avoid the searchlights and anti-aircraft guns around the port of Dieppe.

Long-range aircraft such as bombers were able to

navigate using networks of directional radio beams. But night fighters carried no advanced navigational receivers, leaving Davey to rely on a simple compass bearing, and any features he could identify on the ground while skimming over at 280 kph in darkness.

When they reached the English Channel, Davey cut the altitude to 125 metres. He hadn't dared retract the undercarriage in case it didn't come out again, and he got Marc and Noah to shine a torch out of the cockpit in an attempt to inspect the damage.

With the British coast in sight they encountered three Hurricanes on a routine patrol. They'd probably been sent to investigate a signal picked up by powerful ground radar stations dotted along England's southern coast.

Every British pilot in the sky that night should have been warned to look out for a lone JU-88 night fighter, flown by a friendly pilot. But Squadron Leader Davey had given enough pre-flight briefings to bored and exhausted pilots to know that this far from guaranteed their safety.

Luftwaffe and RAF planes used different radio frequencies, so Davey had no way of talking to his fellow RAF pilots. His only way of showing friendly intentions was to put on his landing lights and then go into a gentle upwards climb.

In this position, the bottom of the aircraft was exposed, making the largest possible target. It was the aviation equivalent of a dog rolling over to let you tickle its tummy.

'Oh Christ,' Davey shouted, as he watched one of the Hurricanes break formation and dive into an attack run.

He threw the JU-88 to one side, but in this position it would only take one machine gun blast to blow their fuel tanks.

Fortunately, at least one Hurricane pilot had paid attention during his briefing – or maybe just realised something was fishy about an enemy plane that flew at you belly up with its landing lights switched on.

Either way, the attacking Hurricane broke off and as Davey levelled out a second Hurricane moved alongside. They were close enough to see goggled faces illuminated by instrument panels and the two RAF pilots exchanged thumbs up.

Two of the Hurricane's resumed their patrol, but to prevent further mix-ups one stayed on the tail of the German night fighter. For the next several minutes, Davey flew calmly, but kept an increasingly wary eye on his fuel gauge.

He studied his instruments. Having the undercarriage down created extra drag, but that wasn't enough to explain the rate at which they'd been burning fuel.

Marc had noticed Davey's increasingly anxious movements, but decided it was better to let his pilot concentrate than ask what was going on and risk breaking his concentration.

Two near-simultaneous events caused Davey to break his silence.

'I've left the choke out,' he shouted, first of all. 'That's a rookie's mistake, damn and blast!'

Marc had learned about choke when he'd learned to drive cars and motorbikes. By pulling out the choke lever, an engine got fed a richer fuel mixture that was needed while it warmed up. But if you left the choke out once the engines were warm, you burned too much fuel.

Then Davey said, 'And that's Worthing blasted pier. Which means we're a good ten miles off course and running on fumes.'

'Can we make it to the airfield?' Marc asked.

'It's touch and go,' Davey said.

Marc and Noah exchanged anxious glances. Neither of them knew enough about flying to discern whether Davey's mistakes were down to incompetence, or just the fact that he was doing a very difficult job. All they could do was keep quiet and let him concentrate.

The other pilot in the Hurricane was confused by the sudden change in course when they reached the coast, but he kept flying just behind. Then the right engine went into a death spiral, choking with several misfires as the propeller slowed to a halt.

'We're too low to glide in,' Davey shouted, as he rolled the plane to the left, hoping to tip any last dregs of fuel towards the surviving engine. 'We're less than ten minutes from Bexhill, but we'll only get a few seconds if the second engine goes, so if I see a good spot I'm going to try and plant her.'

The plane felt twitchy with only one engine running and Marc's heart thudded when the engine spluttered, but mercifully kept running.

'I'm seeing a good flat stretch of field up ahead,' Davey shouted, as he turned the plane gently.

The left engine stuttered and misfired twice more.

'You two brace yourselves. It's going to get bumpy.'

The plane landed hard, with grass and stones pelting the underside, as Davey used left and right flaps to try keeping the JU-88 balanced on its one good wheel. All was good until their only wheel ran into the stump of a felled tree.

The wheel tore off, along with the engine pod under the right wing. The frame of the aircraft buckled, making the cockpit canopy break its bolts and shoot open. Marc's neck snapped painfully as Noah's huge bulk crushed him against the side of the plane.

They were pirouetting. The sound was deafening and a boulder dented the fuselage right next to Marc's head.

As the speed decreased, the rustling and ripping sounds grew more like normal. It was almost a relief, but in a final act the right wing broke off completely and they tilted tail first into a narrow stream before coming to a complete halt.

There was only moonlight. Noah had thumped his head on something hard and moaned as Marc struggled to get out from beneath him.

'Noah, move,' Marc shouted, as he felt around in the dark, trying to release his safety harness.

The big Canadian was concussed, but conscious. Up front, Davey had smashed his face on the control stick and he was spark out, with his nose caved and a deep gash in his cheek.

Even the fumes in an empty tank of aviation fuel are enough to cause an explosion. Marc fought his seat buckle and used every bit of strength to push Noah to one side and knock off the remains of the shattered cockpit.

'What's going on?' Noah asked, completely off his head.

'You need to stand up and get out of the plane,' Marc shouted. 'Can you understand me?'

As Marc stepped out of the cockpit he realised they were at a steep angle. He could step backwards, then slide down the unbroken left wing to the ground, but he didn't want to abandon Noah and Davey.

'Come on, you fat bastard,' Marc said, as he undid Noah's harness and gave him a tug.

Noah almost head-butted Marc as he stood up. After

the noise of the crash, the plane was making eerie sounds. Broken hydraulics hissed and hot engine parts pinged and gurgled.

As Noah slid face-first down the wing, Marc leant over the cockpit sill and reached into a bloody mess to undo Davey's harness. It took everything he had to drag the pilot's torso over the side of the cockpit.

'Stop what you're doing and put your hands where I can see them,' someone shouted, in English.

Marc had been too busy huffing and grunting to hear two men and a woman crossing the field towards the wreckage.

Marc spoke in English, 'I'm trying to get the pilot out. He's RAF!'

A shotgun blast rang out and a man shouted angrily. 'That was a warning shot. Next one won't be, you devious Boche bastard.'

Marc had no choice but to step away from the cockpit and slide down the wing towards a pair of waiting shotguns.

'He's an RAF pilot,' Marc repeated.

'We saw that Hurricane shoot you down,' the younger of the two farmers said irritably.

Marc was pissed off because he wanted to help Davey, but he understood the gunmens' perspective. They'd seen a German plane chased by a Hurricane, followed by a crash landing and the emergence of two people who spoke English with funny accents.

Noah was starting to get his senses back, and looked up at the shotgun pointing in his face.

'We should move further from the aircraft,' Noah said. 'If it catches light we'll know all about it sitting here.'

Marc looked around towards the nose of the aircraft, as Davey groaned from up in the cockpit. At least it meant he was alive.

Although several of the aerial pieces had snapped off, Marc was optimistic because the nose cone with the sensitive radar set inside appeared to have suffered nothing more serious than scuffs and dents.

But the men with the shotgun were on edge. They didn't like Marc looking around, or Noah giving them orders.

'Devious Kraut bastard,' the younger of the two farmers said, before swinging the butt of his shotgun and knocking Marc cold with a blow to the temple.

CHAPTER THIRTY-NINE

CHERUB campus looked different. There was a new perimeter fence and a US Air Force team working security on the front gate. The road up to the old village school where CHERUB agents lived was a sea of mud, churned up by trucks and construction machinery building a runway at the eastern end of what had been a British Army artillery firing range when Marc left a year earlier.

Marc had spent two nights in hospital under observation. He had a dressing over his right temple and mild burns on the back of his neck as he stepped out of Superintendant McAfferty's little Austin.

Inside, a toddler sat beneath a table in the hallway, playing with spent shell cases. Charles Henderson's head popped out of his office and his face lit up when he saw Marc.

'Shit, shit, shit!' Henderson shouted. 'Bloody good to see you!'

'Shit, shit, shit!' the toddler under the table repeated.

'Terence, what have I told you about Daddy's naughty words?' McAfferty asked, as she gave Henderson a dirty look, then picked up the smirking toddler and gave him a good squeeze and a kiss on the cheek.

Henderson and Marc exchanged a solid British handshake, followed by a Gallic exchange of kisses on the cheek.

'Seems quiet,' Marc said, as he looked up the staircase towards the rooms where young agents and trainees slept.

'If they're not away on a mission, they're out training,' Henderson explained. 'I'll need to give you a full debrief on what you've been up to, but from what I've heard so far you've put on a top show. The RAF is chuffed with their radar set. I hear the boffins have already got it working.'

McAfferty spoke firmly. 'Marc's still groggy from the concussion, Captain. He's been away for a year, I'm sure your debrief can wait for a day or two. Right now he needs rest.'

'Absolutely, take your time!' Henderson said cheerfully. 'Head upstairs. Come down for a chat when you feel ready for it.'

As Marc headed upstairs, the familiar smells of floor

polish and steam from the showers became a warm reminder of his friends and his training. His room was a former classroom, at the end of a first-floor corridor which he shared with five other agents.

Paul Clarke was the only person in the room. He lay on his bed, sketching on a small pad. He'd grown ten centimetres while Marc had been away, and had one ankle heavily strapped.

'Still finding excuses to get out of training then?' Marc said brightly.

Paul smiled, then put down his pad and stood up to clear a load of clothes and junk that had built up on Marc's bed while he'd been away.

'We weren't expecting you until this evening,' Paul explained. 'Sounds like you had a good little holiday.'

Marc laughed. 'Nice weather, nice people. It seemed rude to hurry back.'

'We thought you were dead,' Paul said.

'A lot of people did. What are you drawing?'

Paul had a major artistic gift, and Marc was impressed as Paul turned his pad over, revealing a surreal image of two topless girls being chased by a giant serpent which had swastikas instead of scales.

'Very attractive,' Marc said, smirking. 'I take it from the large breasts that you've grown out of the *girls are yucky* phase.'

'Must have,' Paul agreed.

'Much else going on while I was away?' Marc asked.

'We nicked some sugar and yeast from the yanks. PT's been brewing beer and selling it to the locals. Groups B and C are fully trained, but they've decided to cap our unit at twenty agents. A few people have been on missions. Henderson says we're riding our luck: no casualties or anything.'

'Nice,' Marc said, as he sat on the edge of his bed.

He'd abandoned almost everything in France in order to fit into the plane, and apart from the boots his black commando gear and weapons had all been returned to British Army stores. There were just three items in the laundry pouch that McAfferty had brought to the hospital to collect his belongings in.

Marc placed his throwing knife and the dead German soldier's pocket watch on his bedside table, then took out the knotted strands of Jae's hair and gave them an experimental sniff. His heart surged as he caught the smell of the farm, mixed with the barest hint of Jae herself.

Marc turned his head so that Paul didn't see his glazed eye. He was finally home, but it didn't feel like a triumph. There was a huge hole where Jae belonged and Marc knew he wouldn't fill it until he walked back through the gates of Morel's farm and pulled his girl up close for a kiss.

EAGLE DAY

Robert Muchamore

Charles Henderson is the last British spy left in occupied France in 1940. He and his four young agents are playing a dangerous game: translating for the German high command and sending information back to Britain about Nazi plans to invade England.

Their lives are on the line, but the stakes couldn't possibly be higher.

Book 2 – OUT NOW

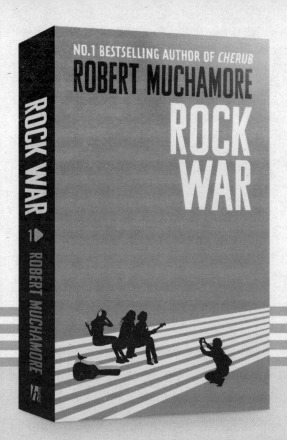

MEET JAY. SUMMER. AND DYLAN.

JAY plays guitar, writes songs for his band and dreams of being a rock star. But seven siblings and a rubbish drummer are standing in his way.

SUMMER has a one-in-a-million voice, but caring for her nan and struggling for money make singing the last thing on her mind.

DYLAN'S got talent, but effort's not his thing ...

These kids are about to enter the biggest battle of their lives. And they've got everything to play for.

ROCKWAR.COM